# UFO's—A SCIENTIFIC DEBATE

CARL SAGAN is Professor of Astronomy and Director of the Laboratory for Planetary Studies at Cornell University.

THORNTON PAGE, Astrophysicist at the Naval Research Laboratory, is now working at the NASA Manned Spacecraft Center in Houston.

# UFO's—
# A SCIENTIFIC DEBATE

EDITED BY

CARL SAGAN and
THORNTON PAGE

The Norton Library
W • W • NORTON & COMPANY • INC • NEW YORK

First published in the Norton Library 1974
by arrangement with Cornell University Press



Published simultaneously in Canada
by George J. McLeod Limited, Toronto

Books That Live
The Norton imprint on a book means that in the publisher's estimation it is a book not for a single season but for the years.
W. W. Norton & Company, Inc.

Library of Congress Cataloging in Publication Data
Main entry under title:
UFO's—a scientific debate.
Reprint of the ed. published by Cornell University Press, Ithaca, N.Y.
Papers presented at a symposium sponsored by the American Association for the Advancement of Science, held in Boston on Dec. 26–27, 1969.
Includes bibliographies.
1. Flying saucers—Congresses. I. Sagan, Carl, 1934– ed. II. Page, Thornton, ed.
III. American Association for the Advancement of Science.
TL789.A1U23 1974 001.9'4 74-10560
ISBN 0-393-00739-1

Printed in the United States of America
3 4 5 6 7 8 9 0

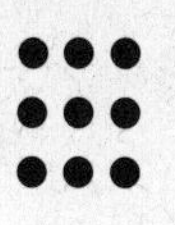

# Contents

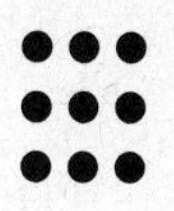

# Illustrations

# Tables

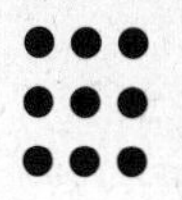

# Contributors

A symposium on Unidentified Flying Objects, sponsored by the American Association for the Advancement of Science, was held in Boston, Massachusetts, on December 26 and 27, 1969. The following participants contributed their papers to this book. A number of these papers have been considerably revised.

Robert M. L. Baker, Jr., Senior Scientist, Computer Sciences Corporation, and Lecturer, School of Engineering and Applied Sciences, University of California at Los Angeles, Los Angeles, California 90024.

Frank D. Drake, Professor of Astronomy, and Director, National Astronomy and Ionosphere Center, Cornell University, Ithaca, New York 14850.

Lester Grinspoon, Associate Clinical Professor of Psychiatry, Harvard Medical School, and Director of Psychiatry (Research), Massachusetts Mental Health Center, Boston, Massachusetts 02115.

Robert L. Hall, Professor of Sociology, University of Illinois at Chicago Circle, Chicago, Illinois 60680.

Kenneth R. Hardy, Chief, Weather Radar Branch, Meteorology Laboratory, Air Force Cambridge Research Laboratories, Bedford, Massachusetts 01730.

William K. Hartmann, Assistant Professor, Lunar and Planetary Laboratory, University of Arizona; presently Senior Scientist, Planetary Science Institute, Tucson, Arizona 85704.

J. Allen Hynek, Professor and Chairman, Department of Astronomy, and Director, Lindheimer Astronomical Research Center, Northwestern University, Evanston, Illinois 60201.

James E. McDonald,* Professor of Atmospheric Sciences, and Senior Physicist, Institute for Atmospheric Physics, University of Arizona, Tucson, Arizona 85721.

Donald H. Menzel, Paine Professor of Practical Astronomy and Professor of Astrophysics, Emeritus, Harvard University, and former Director, Harvard College Observatory; Senior Scientist Emeritus, Smithsonian Astrophysical Observatory, Cambridge, Massachusetts 02138.

Philip Morrison, Professor of Physics, Massachusetts Institute of Technology, Cambridge, Massachusetts 02139.

Thornton Page, Professor of Astronomy, Wesleyan University, and Research Associate, NASA Manned Spacecraft Center, Houston, Texas 77058.†

Alan D. Persky, Associate in Medicine (Psychiatry), Peter Bent Brigham Hospital; Consultant in Psychiatry, Massachusetts Mental Health Center; Clinical Instructor in Psychiatry, Harvard Medical School, Boston, Massachusetts 02115.

Douglass R. Price-Williams, Professor of Psychology, Departments of Psychiatry and Anthropology, University of California at Los Angeles, Los Angeles, California 90024.

Franklin Roach, Affiliate Astronomer, University of Hawaii, Honolulu, Hawaii 96822.

Carl Sagan, Professor of Astronomy, and Director, Laboratory for Planetary Studies, Cornell University, Ithaca, New York 14850.

Walter Sullivan, Science Editor, *New York Times,* New York, New York 10036.

* Deceased.

† Since 1971, Research Astrophysicist, Naval Research Laboratory, Washington, D.C. 20390.

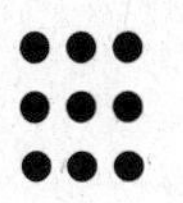

# Editors' Introduction

What does an average person do when he sees something strange, inexplicable, or haunting? How does a scientist react when confronted with observations he cannot readily classify—observations that may challenge some of his most deeply held beliefs? What happens when the observations appear to bear on questions of national security, so that a branch of the armed forces is called upon to perform investigations of a phenomenon it does not understand? What happens when an attempt is made to confront diverse points of view about these events? This book explores what happens. The controversy over unidentified flying objects has for more than twenty years been a lively topic, with representatives of the United States Air Force, the scientific community, and interested public organizations frequently at odds one with another. The public interest in the subject, but only a little of the scientific interest, derives from the idea that unidentified flying objects are space vehicles sent to the earth from elsewhere in the universe. Unambiguous evidence of the extraterrestrial origin of UFO's is obviously not at hand; otherwise there would be no basis for disagreement. Primarily in response to public interest in the topic, the Air Force sponsored a two-year study directed by Professor E. U. Condon at the University of Colorado; the results were published as *Scientific Study of Unidentified Flying Objects* (New York: Bantam Books, 1969), usually referred to as the Colorado Report or the Condon Report.

In the year preceding the publication of the Condon Report, the editors of this book approached the American Association for the Advancement of Science with the idea of organizing a general symposium at an annual meeting of the Association to discuss the UFO issue. The AAAS Board approved such a symposium for the December 1968 meeting in Dallas, Texas. But, largely because the Condon Report—one of the most detailed examinations of the subject ever performed—would not have been published in time to be digested for the Dallas meeting, and in part because of opposition from some scientists, the symposium was postponed for a year and was finally held on December 26 and 27, 1969, at the annual meeting of the American Association for the Advancement of Science in Boston. The organizing committee for the symposium at this meeting consisted of Philip Morrison, Department of Physics, Massachusetts Institute of Technology; Walter Orr Roberts, University Corporation for Atmospheric Research, Boulder, Colorado; Carl Sagan, Laboratory for Planetary Studies, Cornell University; and Thornton Page (chairman of the AAAS special committee), Department of Astronomy, Wesleyan University. The symposium could not have been held without the steadfast courage (sometimes in the face of very heated opposition) of the American Association for the Advancement of Science and, in particular, of the Association's then president, Professor Roberts.

The topic of the symposium is, to the surprise of no one, controversial. The AAAS special committee spent a year and a half arranging a program that attempted to present as fairly and as logically as possible the facts and alternative interpretations that have been offered. The present volume, an only slightly compressed version of the fifteen invited talks and the discussion that followed, is not intended to establish any one interpretation as the "correct" one, but rather to offer the observations and some of the speculations generated by a critical examination of the evidence—the traditional scientific method. Scientists, being human beings, do not always approach controversial subjects dispassionately, and the reader will occasionally find in these pages the heat of passion as well as the light of scientific inquiry.

The opposition to holding this symposium, presented in part by some very distinguished scientists, was based upon the view that if such an unscientific subject as the UFO controversy is discussed, we might just as well organize symposia on astrology, the ideas of Immanuel Velikovsky, and so forth. We believe this conclusion is substantially correct, but it is not the *reductio ad absurdum* that its authors seem to believe it is.*

All of us who teach at colleges and universities are aware of a drift away from science. Some of the most sensitive, intelligent, and concerned young people are finding science increasingly less attractive and less relevant to their problems than was the case for previous generations. We all agree that this drift is deplorable. It must be due in part to their misunderstanding of what science is about, the scientists' failure to communicate its power and beauty. At the same time there is a range of borderline subjects that have high popularity among these same people —including UFO's, astrology, and the writings of Velikovsky. We believe that part of the reason for this popularity is precisely that they are often beyond the pale of established science, that they often outrage conservative scientists, and that they seem to deny the scientific method. We have only to pick up the *New York Times Book Review* (September 21, 1969, for example) to find books advertised under such rubrics as "Science says it shouldn't work—but can't explain why it does!" or "Long-suppressed by so-called 'orthodox science' " or "Unearths new evidence that shakes smug complacency of Establishment scientists."

But while we may deplore this trend, particularly in its extreme variant as a religious cult, it seems to us unprofitable to ignore it. To talk of "dignifying it by discussing it" is to misunderstand these attitudes. They already are dignified in the sense of having widespread newspaper and magazine coverage which reaches many more Americans, both scientists and laymen, than, for example, the scientific journals that generally avoid such discussion.

There are some things we can expect of scientific symposia on such topics and other things we cannot expect. We will not convert true believers, regardless of the strength of our arguments. One religious sect

* This and the following eleven paragraphs are adapted from a letter by Carl Sagan dated September 29, 1969, and addressed to participants in the symposium, the Board of the American Association for the Advancement of Science, and some other interested parties.

confidently predicted the world would end in 1914. Since the world has apparently not ended, one would expect the membership of this sect to be close to zero. This is not the case; its membership has in fact been monotonically increasing since 1914. But what *can* be done in such symposia is to confront unscientific claims and methods with the power of the scientific method. The habits of critical interrogation and of suspending judgment in the absence of adequate data are unfortunately uncommon in everyday life.

Science has itself become a kind of religion, and many pronouncements cloaked in scientific attire are blandly accepted by much of the public. We believe that organizations like the AAAS have a major obligation to arrange for confrontations on precisely those science-related subjects that catch the public eye. Previously such confrontations have served science well. For example, in the Huxley-Wilberforce confrontation on evolution, the novel position has stood the test of time, but the belief that the asteroid Icarus would impact the earth in 1968 has not stood the test of time. In both cases, science has been served well by demonstrations of its power and predictiveness. Recent meetings of the AAAS have shown salutary trends toward increased public relevance —largely on the application of technology for the public good. There seems to us to be an equally important area which has not been adequately stressed, namely, the application of scientific thinking to problems of human interest. Symposia on such subjects as unidentified flying objects can play a significant role in correcting this omission.

There are other topics that might illustrate the scientific method as well or better; but not all of these are in the public eye. All the speakers in this symposium have made recognized scientific contributions. When there is a difference of opinion between scientists with such established credentials, we believe the scientific community is honor-bound to keep the lines of communication open and to aid constructive discussion. We do not see how such a symposium can fail to serve science well.

A similar position has been stated in the document "Science and the Future," a conference summary of a joint meeting sponsored by the British Association for the Advancement of Science and by the American Association for the Advancement of Science, April 13–19, 1969, in

Boulder, Colorado. In a discussion of "Education throughout Life" the following paragraphs appear:

> Students are often taught "the scientific method" in a rigid and formal way which neglects the role of creativity, which reduces its intellectual and social value, and which implies that it is a limited sequence of steps. It should, instead, be thought of as a continuing series of predictions, tests (with adequate controls), and creative hypotheses, and it can only be thoroughly understood by active involvement in this continuing process. There is danger in mistaken ideas amongst the general public of what constitutes a scientific experiment; many "experiments" are performed by individuals, but few of them are scientific in any sense. It may well be far more important to have a large body of people who know how to choose between alternatives on public policy matters based on science, or at the least to be able to follow complex arguments, than it is to have people understand detailed procedures of scientific methods. Perhaps the Associations should include in their programs doubtfully scientific areas of current public interest, such as astrology, extra-sensory perception, and unidentified flying objects, to show how these can be considered in a scientific way.
>
> It is clear to us that the present and future well being of mankind depends upon scientific knowledge. Distrust of science, however, commonly arises from ignorance, or from a mistaken idea of the motivations of scientists. It is very important for young people to know that a "self-correcting" process is inherent in science. Although scientists are aware of this, young people must learn that science and scientists are not free from error and other limitations. Positive and creative attitudes should be promoted, especially at Association meetings, rather than mere negative or apologetic stands.
>
> We consider that it is desirable to have courses at school level on choice-making, and on the difficulties of making true judgments when one is too close to a subject. The Associations should help in discovering such courses if they exist, in developing them if they do not yet exist, and in any case by promoting continuing discussion and study through symposia and other means, and by expressing publicly their concerns about these matters. A conscious and explicit presentation of value-judgments is needed at all stages, together with statements of what choices are involved and of what possibly different points of view may exist.
>
> What is good at the present time in one field and for one country may be evil for the future, or in another field, or for another country or for the

world. Decisions involving value-judgments must be made, and we should stress that the avoidance of decision is in itself a decision.*

The statement was authored by a subgroup chaired by Kenneth Hutton of the British Association, with William Kabisch of the AAAS as Rapporteur and the following scientists as members: E. U. Condon, Ian Cox, Steven Dedijer, Dame Kathleen Lonsdale, Robert Morison, and Carl Sagan.

The order of presentation of topics in this book has been slightly altered from that in the original symposium for reasons of coherence. The first section is introductory and historical; the second is largely devoted to the observations and debates on their interpretation; the third, to social and psychological aspects of the UFO problem; and the fourth is a retrospective and perspective. The participants are astronomers, physicists, sociologists, psychologists, psychiatrists, and a representative of the communications media, and they include most of the scientists primarily involved in discussion of UFO's over the years.

This book begins with a collection of UFO cases, including some of the instructive, puzzling, or fashionable ones discussed from various points of view later in the book. The first paper, by Thornton Page, argues that while much of the UFO reporting has been abysmally sensational and inaccurate, widespread public interest in the subject, viable to the present day, makes the discussion of UFO's an ideal medium for introducing many related scientific issues. William Hartmann and Franklin Roach, both professional astronomers who participated in the preparation of the Condon Report, present different perspectives on the UFO problem. Dr. Hartmann discusses a range of misapprehended natural phenomena that have been identified as UFO's, considers the photographic evidence on UFO's which he has studied extensively, and concludes that "there may be fewer than a dozen [cases] that involve phenomena marginally outside the borders of accepted science." Dr. Roach argues that the cumulative sky coverage of astronomical telescopes is so small that no strong negative conclusions can be drawn from the fact that professional astronomers have not reported UFO's.

* "Science and the Future," June 1969, Appendix 2, American Association for the Advancement of Science.

He suggests that there may be large numbers of civilizations substantially in advance of our own which are capable of interstellar space flight.

J. Allen Hynek summarizes his twenty-one years of experience with UFO reports as principal scientific consultant to the U.S. Air Force Project Bluebook. After "detailed examination of thousands of reports and interrogation of hundreds of witnesses" Dr. Hynek concludes that the unexplained cases "do not specify any known physical event . . . [or] any known psychological event or process." He believes that puzzling cases have been explained away in much too cavalier a fashion, that there is a convention of ridicule which prevents interesting cases from seeing the light of day, and that the UFO phenomenon demands serious scientific attention. He is cautious on the question of explaining the unresolved cases.

The late James McDonald was extremely critical of U.S. Air Force handling of the UFO problem but stressed that the inadequacies of Air Force investigations were due to incompetence rather than conspiracy. Charging inadequacy in all past UFO investigations, McDonald writes: "I speak not only from intimate knowledge of the past investigations, but also from three years of detailed personal research, involving interviews with more than five hundred witnesses in selected UFO cases, chiefly in the United States. In my opinion, the UFO problem, far from being the 'nonsense problem' it has been labeled by many scientists, constitutes an area of extraordinary scientific interest." And he concludes: "It is difficult for me to see any reasonable alternative to the hypothesis that something in the nature of extraterrestrial devices engaged in something in the nature of surveillance lies at the heart of the UFO problem."

Donald Menzel, author of one of the earliest books on flying saucers and also a consultant to the Air Force Project Bluebook, contends that all of the UFO reports can be understood in terms of misapprehended natural phenomena. Some of these, he argues, may be very complex—although most explanations turn out to be fundamentally simple. He uses meteorological, astronomical, biological, human technological, and psychological explanations for many puzzling cases and is critical of those who maintain there is a residuum of inexplicable cases. His report includes seven appendixes, four on particularly interesting cases, one on whether flying saucers tend to move in straight lines, and two on UFO's in art and in the Bible.

Kenneth Hardy draws upon his experience with ultrasensitive radar systems at the Air Force Cambridge Research Laboratories. He describes how radar echoes can be caused by atmospheric refraction and offers remarkable cases of detection of single birds or insects over distances of many miles. R. M. L. Baker, Jr., describes his detailed analyses of four motion pictures of UFO's which have features not completely explained, in his view, as well as about a dozen other films which he concludes were either hoaxes or photographs of natural phenomena.

Initiating the discussion of the social and psychological aspects of the UFO problem, Robert Hall discusses rumor processes, belief systems, mass hysteria, hysterical contagion, and systematic misperception. He believes, however, that some of the "hard-core UFO reports stand up better than many a court case" as far as witness credibility is concerned, and contends instead that scientists may have been unwilling to accept evidence that threatens *their* belief systems. Douglass Price-Williams is interested in the epistemology of UFO reports and the extent to which the actual nature of the phenomenon is hidden by the attempt of the observer to describe it. He believes that a massive statistical search for invariants across a mass of puzzling reports may reveal the intrinsic nature of some of the puzzling reports.

Lester Grinspoon and Alan Persky, drawing upon a background in psychiatry, are impressed by the unusual emotions exhibited by both witnesses and interpreters of UFO phenomena (including at least some of the participants in the symposium). Unlike other topics in scientific study, UFO's seem to rouse a fervor usually reserved for politics, morality, or religion. Drs. Grinspoon and Persky discuss a range of unconscious mental processes, arguing that a large and possibly growing fraction of the community, while generally normal, is subject to transient mental disturbances in stressful situations. UFO's provide powerful and universal psychological symbols.

In a lucid discussion of the reliability of witnesses of scientific phenomena, Frank Drake chooses three examples: (1) a case of seemingly clear-cut photographic evidence for a flying metallic disk; (2) the apparently verified report of the receipt of a television signal from a station three years off the air; and (3) conclusions drawn from eyewitness reports of meteor falls. All three examples are from Professor Drake's personal experience and may shed considerable light on phenomena of the UFO type. Walter Sullivan, the science editor of the *New York Times,* describes the connection between the UFO phenomenon and the news media. He discusses the subtle impediments to more precise reporting and holds that at the very least the UFO phenomenon provides a useful case study of the processes that tend to mold our belief systems.

In the first of two retrospective papers, Carl Sagan argues that there is insufficient evidence to exclude the possibility that some UFO's are space vehicles from advanced extraterrestrial civilizations, but he maintains that other speculative hypotheses are equally probable or improbable, and that the insignificance of our civilization and the vast distances between the stars make the extraterrestrial hypothesis unlikely. He discusses some of the psychological factors molding popular beliefs in UFO's. Philip Morrison's summary of what constitutes scientific evidence is extremely rich in concrete cases connected with UFO's and is a model of critical analysis of controversial phenomena.

In his remarks which opened the symposium, Professor Roberts stressed his belief that the public understanding of science is at stake and that the borders between scientific and nonscientific discussion need explicit delineation. He expressed his hope that the discussion would be well balanced and provide that self-correcting process required for the advancement of science. We hope that the present volume contributes toward this goal.

CARL SAGAN
THORNTON PAGE

## ACKNOWLEDGMENTS

The American Association for the Advancement of Science made the tapes of the three three-hour sessions (chaired by Professors Roberts, Sagan, and Page) available to us; we appreciate their cooperation.

We are particularly grateful to Mrs. Mary Szymanski for skillful extraction of the texts of three full papers from noisy magnetic tape recordings. The present volume would not have been possible without her help. Our thanks also to Mrs. Marye Wanlass for retyping much of the text of this book, and to Miss Lucy Ahn and to the staff of Cornell University Press for helping to see this volume through the final stages of publication. The index was prepared by Theodore Szymanski.

Some of the material in Chapter 4 also appears in *The UFO Experience* by J. Allen Hynek (Chicago: Regnery, 1972). We acknowledge with thanks those publishers and individuals who gave us permission to reproduce illustrations and other material, including: the Institute of Electrical and Electronic Engineers for permission to reproduce two figures from K. R. Hardy and I. Katz, in *Proc. IEEE,* April 1969, pp. 470, 471; Bantam Books, Inc., for permission to adapt Figures 3 and 4 from *Scientific Study of Unidentified Flying Objects* by Dr. Edward U. Condon (Copyright © 1968 by The Board of Regents of the University of Colorado); *Flying Saucer Review* for permission to print, in revised form, Donald M. Menzel's article which appears as his Appendix 5; the American Association for the Advancement of Science, for permission to reprint Appendix 2 of "Science and the Future," June 1969; Dr. David Atlas for permission to reproduce portions of a letter written to Carl Sagan, 29 September, 1969; Professor Thomas A. Goudge for permission to reproduce parts of his letter to J. Allen Hynek; and the Office of the Secretary of the Air Force for permission to reproduce letters written to Thornton Page and Donald M. Menzel in January, 1970.

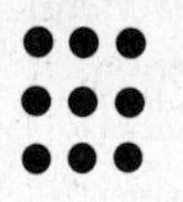

# Selected UFO Cases

In the following brief summaries, each case is identified by the *location, time,* and *type* of sighting. The types of sightings are explained in Chapter 4. Several of the authors refer to these illustrative cases later in this book. These summaries were prepared by Thornton Page.

### Alaska. January 22, 1952, 12:10 A.M. Radar

A strong target appeared on the radarscope, moving in from the northeast, fairly high, at *1,500 miles per hour*. Three jet fighters took off from an airbase 100 miles to the south and were vectored toward the unknown by ground radar but never saw any visible target. When the ground radar was switched to short range, both unknown and fighter planes disappeared from the screen. Two of the fighters picked up a stationary target on their airborne radars over a period of 10 minutes. (Air Force investigation concluded that these were ground-radar returns caused by pecular atmospheric conditions.)

### Artesia, New Mexico. January 16, 1952. Daylight Disk

Two members of a balloon project from the General Mills Aeronautical Research Laboratory and four other civilians observed two unidentified aerial objects in the vicinity of the balloon they were observing. The balloon was at an altitude of 112,000 feet and was 110 feet in diameter at the time of the observations.

The objects were observed twice, once from Artesia, and once from the

Artesia Airport. In the first instance, one object appeared to remain motionless in the vicinity of, but apparently higher than, the balloon. It had twice the angular diameter of the balloon and its color was a dull white. This observation was made by the two General Mills observers.

A short time later the two observers and four civilian pilots were observing the balloon from the Artesia Airport. Two objects at apparently extremely high altitude were noticed coming toward the balloon from the northwest. They circled the balloon and flew off to the northeast. The duration of observation was about 40 seconds, and the two objects were the same color and size as the first object observed from Artesia. The two objects were flying side by side, and when they appeared to circle the balloon, they temporarily disappeared, causing the observers to assume they were disk-shaped and had turned edge-on to bank.

### Colorado Springs, Colorado. May 13, 1967, 8:40 P.M. Radar

The weather was overcast, with scattered rain-and-sleet showers and gusty winds. As a Braniff airliner came in for a landing, the ground radar detected a target beyond it at about twice the range. As the plane landed, this target pulled to the east and passed low over the airport (at 200-feet altitude, about a mile and a half from the control tower). The tower operators, alerted by the radar operation, *saw and heard nothing*. The pilot of another aircraft, 3 miles behind the Braniff plane, saw nothing when asked to look. (The Condon Report, p. 170, calls this "one of the most puzzling radar cases on record.")

### Deadwood, South Dakota, September 22, 1966, approximately 3:00–4:00 A.M. Nocturnal light

On this clear night, stars were all visible, and there was a very light breeze. At about 2:40 A.M., Officers A and B were patrolling highway U.S. 14 to the North end of "76" Hill, a mountain extending up out of the canyon at the north end of Deadwood. Officer A stated that as they drove up to the top of the hill, they noticed a *large, white, round object in the sky,* a little to the northeast of them. They stopped at a parking area on top of the hill and were facing east northeast as they observed the object at about a 50 degree elevation angle, apparently between Deadwood and Sturgis. Officer A radioed to Rapid City on the car's state radio and asked if they could see it. Rapid City replied in the negative. Sturgis then radioed Officer A that they could see the object in the direction of Deadwood, so it apparently was *between Deadwood and Sturgis*. The radio operators at Spearfish, Belle Fourche, and Leed all radioed that they could also see the object. Officer A stated that they watched it hang motionless in the sky for 15 to 20 minutes. On two or three occasions, he shined the police car spotlight on the object; it would black out, then come back on when the spotlight was turned off. Also, during this period, it turned *pale green,* then *red,* then *white.* It was about the size of a silver dollar held out at arm's length. After about 20 minutes, they noticed a *smaller white object*' about the size of a pea held at arm's length, streaking in toward the larger object from the northwest and then stopping. Then another object, the same size, streaked in from the southeast and stopped close to the larger object. Presently the larger object moved to the right, then down, then to the left, then up again, in a square. As it did this, it would send out occasional *blue shafts of light* toward the ground. These shafts of light would last only 2 or 3 seconds, then go out. Again Officer A shined his spotlight on the larger object and it would go out, then come back on when the spot was turned off. The radio operators at the other locations also radioed that they could see this object manuevering, with the other two remaining motionless in a fixed position. After about 30 minutes, the smaller objects shot off at high speed in the directions from which they had come, taking about 5 seconds. For another 25 minutes or so, the larger object stayed in one spot, shooting out shafts of blue light; then it moved at high speed, stopping, backing up, then moving forward again at high speed, until finally it too disappeared into the southeastern skies. *No noise* was detected from any of the objects. *No airplanes* were heard or observed during the sighting.

### Gravois, France, September 22, 1967, about 8:30 P.M. Nocturnal light

A Catholic priest writes: "I was the last person in the world to imagine that I had seen a UFO because I have been very skeptical. . . . I was coming south from Versailles, about 3 miles outside of Gravois, when I noticed this light about the size of a big grapefruit. Because this light was so clear and bright and low lying, I suspected that it was a reflection from some light from the ground reflected in the window, so I turned the window down. The light was still there, so I pulled the car into the side of the road. It appeared not too far away, and I watched it for 15 minutes. There was a constant stream of traffic going toward the lake. It amazed me that no one else got out of their cars, but kept on going. I was standing there on the side of the road looking up. It was dusky. After a while the object seemed to move off, but it didn't move in a uniform motion. It made kind of a round swing and eventually it seemed to head off toward the northwest and then it swung a little to the north and then it seemed to go toward the northeast.

"I was about a mile from another gentlemen, who was equally skeptical, and I told him what I had seen. He and his son had seen the same thing. I saw it definitely stationary for about 15 minutes. I timed it. It seemed to me to be going far faster than an airplane; it couldn't have been an airplane. It was a bright yellow, and had no definite shape except for a while I thought it was kind of flat with a dome shape on top."

### Haneda Air Force Base, Tokyo, Japan, August 5, 1952, 11:30 P.M.–12:30 A.M. Nocturnal light and radar (see also Chapter 5, Appendix, Case 3)

In the clear night sky, several ground observers saw a *bright round light* low in the northeast, and one incoming pilot after radio query said it looked like a bright star. The ground radar could at first find no target. At about 11:50 a fighter plane was vectored in on a bogie, made *contact with airborne radar,* but the pilot saw nothing visually and lost the radar contact in 90 seconds; the target's estimated speed was very high. (The Condon Report, p. 126, identifies the visual sighting as a *star*.)

### Kirtland Air Force Base, Albuquerque, New Mexico, November 4, 1957, 10:45 P.M. (see also Chapter 5, Appendix, Case 4)

Two men were on duty alone in the control tower at Kirtland AFB, New Mexico; the tower is slightly over 100 feet high. One of the controllers looked up to check cloud conditions and noticed a *white light* traveling east at *200 miles per hour* at an altitude of approximately 1,500 feet. He called the radar station and asked for an identification of the object. The radar operator reported that the object was on a 90-degree heading. It angled across the east end of Runway 26 in a southwesterly direction and began a sharp descent. One witness gave a radio call in an attempt to contact what was believed to be an unknown aircraft that had become confused about a landing pattern. The object was then observed through binoculars and appeared to have the shape of "an automobile on end," about 15–18 feet high. One white light was observed at the lower side of the object. The object slowed to fifty miles per hour and disappeared behind a fence one-half mile from the control tower. It reappeared, moving eastward at an altitude of 200–300 feet; it then veered in a southeasterly direction, ascended abruptly at an estimated rate of climb of 45,000 feet per minute, and disappeared.

Although there were scattered clouds with a high overcast, visibility was good. Surface winds were variable at ten to thirty knots. Witnesses observed the object for 5 or 6 minutes, approximately half of that time through binoculars.

The radar operator stated that the object was first sighted near the east boundary of Kirtland AFB. *It reversed course and proceeded* to the Kirtland low-frequency-range station where it began to orbit, then left at high speed and disappeared 10 miles from the observer. About 20 minutes later an AF C-46 took off to the west. The observer scanned radar to the south and saw the object 4 miles south. It made an abrupt turn to the west and fell into trail formation with the C-46. The object maintained approximately one-half-mile separation from the C-46 for approximately 14 miles, then hovered for approximately a minute and a half, and faded from the scope.

### Lakenheath, England, August 13, 1956, 11:00 P.M.–3:30 A.M. Radar and nocturnal light (see also Chapter 5, Appendix, Case 2)

Two RAF ground-radar stations detected several objects moving at high speed on a clear moonlit night. The first radar tracked one traveling at about *3,000 miles per hour* westward at *4,000 feet altitude;* simultaneously, tower operators reported a bright light passing overhead toward the west and the pilot of a C-47 aircraft at 4,000 feet over the airfield saw the bright light streak westward underneath him. The second radar station, alerted by the first, detected a stationary target at about *20,000 feet altitude* that suddenly went north at *600 miles per hour.* It made several sudden stops and turns. After 30 minutes an *RAF fighter* was called in and made airborne-radar contacts with the object over Bedford (just north of Cambridge, England). Suddenly the object moved around behind the fighter plane, both being tracked by ground radar. The fighter pilot could not "shake" the object. A second plane was called in but never made contact and all radar contacts were lost. Several other radar targets were tracked in the same area and several other small moving lights were seen; all disappeared at 3:30 A.M., by which time a few clouds appeared in the sky.

### Methuen, Massachusetts. January 20, 1967, after dark. Close encounter

Three people were driving northeast through a lonely area bordered by woods, field, and a few houses. Reaching the top of the hill they suddenly came upon a straight *string of bright glowing red lights* moving northeast along the roadside to the north. They appeared to be at an altitude of 500–600 feet and just off the road at a point estimated to be about 400–500 feet away from them. Witnesses immediately slowed the car and proceeded toward the lights. When almost broadside to the lights which now seemed to be hovering, the object to which they were apparently attached swung around in a smooth sideways turn revealing a new light configuration and color. Four distinct lights formed a perfect trapezoid: two red lights formed the top and two white lights formed the base. One witness was certain she saw a dimmer white light just above the two red lights. All were impressed by the large size of the individual lights and the apparent size of the object that they must have been attached to. The red lights were compared to the color and brightness of a hot electric stove burner. A reflecting metal was seen behind the lights. The center of the trapezoid seemed to be dark and nonreflecting. The driver pulled over to the side of the road directly broadside to the object which seemed to be lower and only 100–300 feet away. The witnesses decided it would be best to stay in the car, which was idling with lights and radio on. Then abruptly the *engine, lights, and radio failed completely* except for the generator light, which just barely lit up and was pulsating off and on. The driver immediately tried to start the car but the engine would only "moan" and would not start. Thinking that the lights and radio might be overloading the battery, the driver tried to start the car again after switching them off but was unsuccessful. The driver had opened the side window. The others were afraid to put down the larger windows. *No noise* was heard. Then the object began moving slowly and then shot away at great speed in a southwesterly direction. The driver was then able to start the car and the lights worked perfectly as did the radio later when they turned it on.

### A small town in Minnesota. June 1958, 6:30 P.M. Daylight disk

"My wife and I had just finished supper. I went out and started the garden hose. The sun had dropped below the horizon but the western sky was quite golden after the rain shower. There was a large thunderhead cloud in the southwest sky. I heard a sort of whining noise and I thought that one of my neighbors about a block away might be running a saw.

"The sound became stronger and a steady whine, and seemed to be coming from the southwestern part of the sky, not like what I had heard of jets. I turned around and looked up toward the thunderhead from where the sound now seemed to be coming. As I looked I saw this thing come out from behind the thunderhead. My wife just saw the last part of it as it went back behind the thunderhead.

"I remarked to her that this was something new in flying machines—probably some new government test.

"There was no mention of anything in our local paper, but later I realized that in town with trees in the streets, the object probably had not been visible. I made a pencil sketch of it; it was near enough so that I got a good view in several positions. It sort of spiraled and glided and was *silvery* with what appeared to be *portholes showing dark* as interiors would. I would judge it to be about 150 feet in diameter. I don't know what height thunderheads usually are but the distance could be judged from that."

Missouri. March 6, 1966, 11:00 A.M.
Close encounter

The sky was clear and the sun was behind the observer, who was driving with her dog, a St. Bernard, sleeping in the back seat. The dog started acting very strangely, barking and seemingly quite upset. The dog jumped up on the front seat with his hair standing up on the back of his neck. Suddenly he acted as though someone had whipped him and tried to get down under the seat. He was whimpering and very scared. The observer then saw a beam of light on the road ahead of the car. The *light beam* extended about one foot over each side of the road, which has a twenty-four-foot pavement, and the beam was *blue-white* in color and bright enough so that the observer could see what appeared to be dust particles in the beam. As the observer looked through the beam the road beyond seemed distorted as though by heat waves. As the car entered the beam it slowed from 50 miles per hour to about 10 miles per hour. As the car began to slow, the observer looked out and up through the windshield and saw a *disk-shaped object* hovering over the road. She estimated it to be some 1,000 feet high; it appeared larger than a dime held at arm's length, to be metal with a raised or domed area at the top. The witness could see no detail, lights on the object, or seams. The light beam narrowed to a small area in the lower center of the disk. The object appeared to be stable; it did not wobble. The surface seemed to be very smooth. The light beam was very bright and the witness had to close her eyes partially to look at the object. She stated that her eyes bothered her for 3 days following the sighting. When the *car slowed* to about 10 miles per hour, she pushed the accelerator to the floor, but the car would not respond. After passing through the beam the automobile ran smoothly again. She then drove straight home and did not look again at the object. The total duration of the sighting was about 10 seconds.

Montgomery, Alabama. July 24, 1948, 2:45 A.M.
Nocturnal light

Pilot C. S. Chiles and co-pilot J. B. Whitted in the cockpit of an Eastern Airlines DC-3 at 5,000 feet altitude en route from Houston to

Boston saw a *dull-red object* approaching on a collision course. During the next 10 seconds, it veered slightly to the right, passed the plane on the right at high speed, then seemed to pull up, and disappeared in the clouds overhead. One passenger on the right side of the plane glimpsed the bright light as it flashed by. There was no disturbance of the DC-3. The pilots described the object as cigar-shaped, about 100 feet long, with two rows of lighted windows, a dark blue glow underneath, and a *red-orange jet flame* about 50 feet long behind it. They estimated the closest approach to be less than a mile. (Both Hynek and Menzel identify this as a *meteor* much farther away.)

### Newton, Illinois. October 10, 1966, 5:20 P.M.
### Daylight disk

A woman and five children witnessed the slow passage of a metallic object past their farm home. Observing conditions were excellent, with clear, dry weather. The object was first seen by the children, ages four through nine years. The mother responded to the children's call and joined them in the yard, walking parallel with the object's motion. The object moved slowly and uniformly in a westerly direction, at walking speed, approximately 50 feet above the ground. The object disappeared by abruptly turning nose up and moving upward extremely rapidly, disappearing from sight in one or two seconds. An analysis of sighting and landmark positions and angular clues suggests a prolate spheroid approximately 20 feet long and 8 feet in diameter. The surface was metallic, like aluminum; the witness was near enough to ovserve longitudinal seams. The object had a small dorsal fin at the rear and a rectangular black aperture near the front. A brownish gold design was observed on the lower rear portion. The whole object was at all times surrounded by a bluish haze about 5 feet thick. The haze had a noticeable optical thickness. It also contained luminous bubbles or sparks. No sound was heard from the object except for an unusual vibrating noise perceived for a few seconds when the object was nearest. No electrical effects were noted by the observers.

The children were asked to sketch the object the evening of the sighting. Individual judgments of color obtained by means of a Nickerson color fan produced consistent results.

Seventy minutes after this observation, under dark sky conditions, an elliptical blue light of the same color and axial ratio was seen moving in the same general direction, at low elevation angle, by a witness seven miles from the location of the first sighting. These two sightings were the first ever reported from this area.

### Salt Lake City, Utah. October 2, 1961, 12:05 P.M.
### Daylight disk

A civilian pilot, taking off from the Utah Central Airport, noticed a bright *silvery disk* ahead of his plane. A few minutes later he saw that the object was *pencil-shaped* and still in the same position. He radioed the control tower, where the operator saw the same object directly under the sun apparently *hovering* over Provo, 40 miles to the south. The pilot flew toward the object, which seemed to be at about 7,000 feet altitude and rocking gently. When he got about 5 miles from it, the object suddenly shot up and retreated rapidly southward soundlessly and with no vapor trail. After a few seconds, while it diminished in size, the object vanished at an estimated speed of several thousand miles per hour. Ground observers at the airport saw the object, but others at Provo, alerted by radio, did not. The sky was slightly hazy (Menzel identifies the "object" as a *sundog* produced by scattering of sunlight in cirrus clouds).

### South-central U.S., July 17, 1957, early morning.
### Radar (see also Chapter 5, Appendix, Case 1)

A U.S. Air Force B-47, especially equipped with *electronic countermeasures equipment* and carrying a crew of six, returned from a test mission over the Gulf of Mexico and was headed north flying at about *30,000 feet*. The weather was clear and all crew members were monitoring their equipment. One special radar, at 2,800-megacycle frequency, detected a strong target overtaking the B-47, and shortly the pilot and co-pilot saw a *bright white light* moving ahead of them. This object veered across the B-47 course toward the east at a speed much higher than any aircraft. It seemed to be "as large as a barn" and was picked up by the 2,800-megacycle radar on the right side of the B-47. The object was flying at the same speed even though the pilot changed the air speed of the B-47. Ground-based radar confirmed the presence of the object about 10 miles east of the B-47. The object then moved to a position ahead of the B-47 and was seen as a *large red glow.* It stopped and, as the B-47 flew over, *disappeared.* At this time the radar target also disappeared. As the B-47 circled to reapproach the last position, the object *reappeared,* stationary on both radars, at 15,000 feet altitude. As the B-47 approached, it disappeared again at a range of 5 miles.

### White Sands Missile Range, New Mexico, March 2, 1967, 12:25–11:32 P.M. Radar

A driver on Highway 70 near the Apache Summit at 9,000 feet elevation reported *silvery specks* passing overhead from north to south. Two ground radars at Holloman AFB searched the region near Apache Summit, found nothing moving from north to south but got *intermittent targets.* Lighter aircraft searched the area but found no visual or radar targets. (The ground-radar targets are explained in the Condon Report, p. 151, as ground targets and a possible drifting *balloon.*)

# PART I

# BACKGROUND

# 1

## Education and the UFO Phenomenon

THORNTON PAGE

At present it is fair to say that the attitude toward UFO's is highly polarized between the conservative views of a small group of senior physical scientists and the vastly more speculative views of a large part of the American public. This book is directed toward a middle group who want to learn more of the facts about the UFO problem, to hear rational discussion of alternative explanations of peculiar sightings, and to go over some of the sociological interpretations of the very widespread UFO phenomenon. (For instance, I hope the sociologists take note of "Page's Law," that the wave of UFO concern moves eastward around the world, completing one full circuit in about seventeen years.) We are convinced that logical discussion of this widely publicized topic will serve a beneficial educational purpose, both among scientists and the general public.

### Educational Aspects

The aim of the symposium, and hence of this book, is to bring the varied facts on UFO's to the attention of scientists and to show enthusiasts the implications of the very much better organized facts in the physical, biological, and social sciences. Two specific educational aspects are important: the possible harm done to science education by pseudoscientific UFO reports, magazines, and books, and the use of student interest in UFO's to benefit the teaching of science.

It is appropriate to begin with a reference to the *Scientific Study of*

*Unidentified Flying Objects* (New York: Bantam Books, 1969) prepared by E. U. Condon and his thirty-six-member staff at the University of Colorado during 1967 and 1968. The symposium was delayed for a year so that the full content of the Condon Report could be read and digested after its publication in January 1969. The paperback edition contains almost 1,000 pages, including case studies, analyses along the lines of several different scientific disciplines, and a twenty-year historical summary. My own experience goes back to the panel convened by H. P. Robertson in 1953, which issued a much shorter report and classified it "secret." Another panel, which met under the chairmanship of Brian O'Brien in 1966 and included Carl Sagan, made the recommendation which led to the Condon study; the O'Brien report also was not widely read.

The United States Air Force, charged with the responsibility for investigating UFO's, has come to realize that public education is needed to alleviate the "UFO problem." About 90 per cent of the 13,000 reports received by USAF Project Bluebook could have been recognized as normal physical phenomena by persons who had studied elementary astronomy in high school or college. Of course, the press and other mass media have influenced public reaction (see Chapter 13), and there is a natural tendency of the average layman to be intrigued by mysterious or unexpected appearances (see Chapter 11). As we all know, public demand helped to build up a large body of published literature, some of it fallacious (or highly speculative) and of special appeal to readers uneducated in science, particularly youngsters of high-school age.

It would be ridiculous to claim that one two-day symposium or one modest book could come up with the "correct" answer to the UFO question. The Condon Report, which involved more than fifty man-years of study, reached a conclusion (disputed by many) that further study is not worthwhile. As a result, the Air Force recently discontinued the Project Bluebook files, which will be moved from Wright-Patterson AFB to the USAF Archives at Maxwell AFB, Montgomery, Alabama (where access will be controlled by the Office of Information, Department of the Air Force, Washington).

Although the Bluebook files were far from perfect, their termination has disappointed many people, including some of the scientists participating in this symposium. As Condon himself writes (in the Condon Report, Section I, p. 2):

> Scientists are no respecters of authority. Our conclusion "that further UFO studies are not worthwhile" will not be uncritically accepted by them. Nor should it be, nor do we wish it to be. For scientists, it is our hope that the detailed analytical presentation of what we were able to do, and of what we were unable to do, will assist them in deciding whether or not they agree with our conclusions.

## The Harmful Effect of UFO Literature

Condon devoted the last half-page of his "Conclusions and Recommendations" to the "miseducation in our schools which arises from the fact that many children are being . . . encouraged to devote their science study time to the reading of UFO books and magazine articles." There can be no doubt that some of the books and articles on UFO's are unsuitable and misleading, just as popular books on science fiction, astrology, drugs, and sex (equally available) may conceivably be harmful to young readers.

A brief review of Lynn Catoe's UFO bibliography [1] shows 71 books, 28 pamphlets, and 73 magazine articles in English that have been available and widely read in the United States since 1947 (the pamphlets and articles mostly since 1961). I have classified these in eight categories (Table 1–1), ranging from "conservative science" and the "Air Force position" through descriptive "historical reports" to speculative treatments of "extraterrestrial visitors," and "contacts" with them. (Since it is unlikely that old magazine articles have much influence today, the pamphlets and articles before 1961 have been omitted from the table.)

The statistics in Table 1–1 are interesting in themselves: the books show a much wider range in "speculativeness" than the pamphlets and magazine articles. About 10 per cent of the books are extremely conservative, and 20 per cent highly speculative, whereas the pamphlets and magazine articles are mostly of the liberal scientific or historical type, eminently suitable for student reading. If books like Adamski's *Inside the Space Ships* (New York: Abelard-Schuman, 1955) are discounted as

*Table 1-1.* General publications on UFO's in English, 1948–1969

| Category | General books | | | | | | Pamphlets | | Magazine articles | |
|---|---|---|---|---|---|---|---|---|---|---|
| | 1948–59 | | 1960–69 | | Total | | 1961–69 | | 1961–69* | |
| | No. | % | No. | % | No. | % | No. | % | No. | % |
| Conservative science | 1 | 4 | 3 | 6 | 4 | 6 | 1 | 4 | 1 | 1 |
| USAF position | 0 | | 3 | 6 | 3 | 4 | 0 | | 1 | 1 |
| Liberal science | 1 | 4 | 5 | 11 | 6 | 8 | 8 | 29 | 35 | 48 |
| Psychology | 0 | | 2 | 4 | 2 | 3 | 1 | 4 | 3 | 4 |
| Historical report | 8 | 33 | 7 | 15 | 15 | 21 | 13 | 46 | 17 | 23 |
| Extraterrestrial speculation | 7 | 29 | 13 | 28 | 20 | 28 | 2 | 7 | 15 | 20 |
| Speculation on secrecy | 2 | 8 | 5 | 11 | 7 | 10 | 0 | | 0 | |
| Stories of contacts | 5 | 21 | 9 | 19 | 14 | 20 | 3 | 11 | 1 | 1 |
| Totals (100%) | 24 | | 47 | | 71 | | 28 | | 73 | |

* Magazines in English devoted entirely to UFO's are *APRO Bulletin* (formerly *Newsletter*), U.S. bimonthly started in 1952; BUFORA Journal, British quarterly started in 1965; *Fate*, U.S. monthly started in 1948; *Flying Saucer Review*, U.S. bimonthly started in 1954; *Flying Saucers*, U.S. monthly started in 1957; *Saucer News*, British quarterly started in 1953; *UFO Investigator*, U.S. monthly started in 1958.

Articles also appeared in *America, Argosy Bluebook, Giant Comic Book, Macleans, New Republic, Life, Look, Newsweek, New Yorker, Playboy, Saga, Spectator, Nation, Saturday Evening Post, Saturday Review, Time, True, U.S. News and World Report;* and (less available) in *Aero Digest, Airline Pilot, Airman, Bioscience, Bulletin of the Atomic Scientists, Journal of the Optical Society of America, Journal of the British Interplanetary Society, Library Journal, Physics Bulletin, Physics Today, Popular Mechanics, Popular Science, Public Opinion Quarterly, Science, Science Digest, Science and Mechanics, Yale Scientific Magazine.*

science fiction, even the speculative books can be used by an experienced science teacher as "interest rousers." As any teacher knows, student interest in a topic—even if it derives from misconceptions—is better than no interest at all.

## Using UFO's in the Teaching of Science

The reader may be intrigued by how I tested this technique at Wesleyan University in an undergraduate elective course ("Science 101"), designed to interest non-science majors who would otherwise have had no science courses whatsoever. Not all my colleagues on the faculty were enthusiastic over my offering "Flying Saucers" in the fall semester, 1967, even when it was oversubscribed and offered again in the spring of 1968 by student demand.

Along with many other science teachers, I had become frustrated with the diminishing undergraduate interest in physical science (Table 1–2) at a time when space exploration, electronic computers, and nuclear physics seem to me to offer more and more exciting work. The lack of interest (even hostility) has been traced to poor teaching of mathematics and physics in grade schools and high schools, but this scarcely helps to solve the problem of what we should do about a generation of college students who want nothing to do with physical science at a time when more young physicists, engineers, and astronomers are needed.

*Table 1-2.* Undergraduate majors at Wesleyan (in per cent graduating in each field)

| Year | Physical sciences | Biological sciences | Social sciences | Humanities and the arts |
|---|---|---|---|---|
| 1960 | 16 | 10 | 50 | 24 |
| 1961 | 15 | 8 | 55 | 22 |
| 1962 | 13 | 11 | 50 | 26 |
| 1963 | 13 | 11 | 51 | 25 |
| 1964 | 12 | 16 | 48 | 24 |
| 1965 | 11 | 10 | 44 | 35 |
| 1966 | 10 | 6 | 59 | 25 |
| 1967 | 9 | 7 | 53 | 31 |
| 1968 | 8 | 6 | 58 | 28 |
| 1969 | 7 | 4 | 59 | 30 |

Briefly, the one-semester course consisted of two lectures and a discussion session each week, with a two-week reading (and writing) period near the end. We started with a review of UFO reports, then spent five weeks on elementary astronomy—because planets, bright stars, and meteors are so often reported as UFO's. The importance of celestial coordinates and time was stressed for proper reporting of UFO's, and students were interested (or villainous) enough to phone me late in the evening at home to report celestial objects that looked like UFO's. One of my most active evenings was in November 1967, when there was a bright "moondog" (ring around the moon) reported to me by every one of the fifty students in the class.

At this point we shifted to atmospheric physics, and discussed ball lightning, refraction, and aurorae for a week or two. Then we returned to astronomy for discussion of the extraterrestrial hypothesis. The students learned that conditions on other planets of the solar system are perhaps not conducive to intelligent life, and discussed theories of the origin of the solar system and the origin of life (a very popular topic). They learned how stellar distances are measured by parallax, discussed the probability of life on planets of other stars, and recognized the difficult problems of interstellar travel (long distance, and impact with interstellar material at high speed [2]).

Each student spent the two-week reading period writing a term paper on a topic selected from the following list:

The Celestial Sphere
- Coordinates in the Sky
- The Constellations
- The Ptolemaic System
- Distances to Planets, Stars, and Galaxies

The Earth's Atmosphere
- Aurorae and Luminous Clouds
- Meteors and their Trails
- The Ionosphere, Radio, and Radar
- Effects of the Solar Wind

Celestial Mechanics
- History of Planetary Motions
- Evidence for Motions of the Earth
- Complete Description of an Orbit
- Newton *vs* Einstein
- Travel between Stars

Space Probes
- History since 1930
- Launch and Guidance into Orbit
- Design of a Modern Space Probe
- Orbits and Times for Interplanetary Flight
- Purpose of NASA Programs

Moon and Planets
- Surfaces of the Moon and Mars
- Theories of Crater Formation
- Living Conditions on Moon and Planets

Solar System
- Differences between Planets, Comets, and Meteoroids
- Solar Flares and the Solar Wind
- Origin of the Solar System
- Evidence for Life on Other Worlds

Flying Saucers
- History
- Survey of Significant Reports
- Sociological Implications
- Physical Peculiarities
- Reliable Identifications

When the papers were turned in, each was passed to a different student assigned the job of writing a critique. In all but a few cases, these critiques revealed a good grasp of the astronomy and physics involved. The three best papers were published in pamphlet form, and sold well (at 25 cents) in the college bookstore. These three authors appeared on a half-hour television show to explain their views on UFO's, and thus gained firsthand experience of the publicity aspects of this topic. Earlier in the semester, two outside speakers widely recognized for their UFO studies (J. Allen Hynek and Donald H. Menzel) had lectured to the class, and told a little about the publicity difficulties.

I am convinced that the students learned a good deal of astronomy, physics, and biology in the "Flying Saucer" course, although I admit that such a course is not suitable as a standard part of the curriculum and that possibly it might lose its appeal after the novelty wore off. For "lab work," the students learned constellations, spotted an earth-orbiting spacecraft, and looked at bright planets through a small telescope. Several searched for UFO evidence on films taken by one of the sixty-four cameras of the Prairie Network,[3] after a session in which we decided that the astronomical telescopes in use have almost no chance of photographing a UFO passing through the telescope field.[4] On the other hand, the Prairie Network has about 65 per cent coverage of the sky for bright objects over 440,000 square miles in the Midwest—about 0.22 per cent of the earth's surface. A similar Canadian network and the earlier Czech network raise this area coverage to about 0.5 per cent. The network results (negative for UFO's, positive for meteors) are discussed later in the book.

## Conclusions

The general advancement of science depends heavily on public education in science. Most of the significant research today depends on public support (university, foundation, or government financing). It is therefore obvious that students (and older citizens) must be given enough education in science to recognize worthwhile scientific effort. For a number of reasons, many students as well as the general public are interested in UFO's. Teachers should capitalize on this interest in teaching courses of broad appeal; scientists in general should take advantage of public interest in UFO's to correct public misconceptions about science.

## NOTES

1. Lynn E. Catoe, *UFO's and Related Subjects: An Annotated Bibliography,* Library of Congress, AFOSR 68-1656 (Washington, D.C.: U.S. Government Printing Office, 1969).

2. Freeman Dyson, "Interstellar Transport," *Physics Today,* Oct. 1968.

3. R. E. McCrosky and H. Boeschenstein, *The Prairie Meteorite Network,* Smithsonian Astrophysical Observatory Special Report No. 173 (Cambridge, Mass.: 1965).

4. Thornton Page, "Photographic Sky Coverage for the Detection of UFO's," *Science* 160 (1968): 1258.

# 2

## Historical Perspectives: Photos of UFO's

WILLIAM K. HARTMANN

Although I was involved in the University of Colorado UFO project and was a coauthor of the Colorado Report,[1] I can present only my own experience, which has shown that in the UFO business one can trust nothing secondhand. Also, in my experience, the UFO evidence is very poor. In studying photographic cases I went to a number of the UFO enthusiasts and asked for their lists of the best cases. From these I chose the strongest UFO cases to investigate, but even these all fell apart upon close examination. They included such photo-classics as Santa Ana, Vandenberg AFB (Air Force tracking films), Tremonton, Beaver County, and the Fort Belvoir ring (cases 49 to 53 in the Condon Report).

Later I reviewed the UFO literature and found that what are regarded by some as the "classic" cases have been adequately explained by other investigators. Often, what one enthusiast quotes as evidence for extraterrestrial intelligence (ETI) or some other extraordinary phenomenon, another ETI enthusiast concedes as an explained case! Thus there is no agreement on what the evidence is. The current status of some famous cases that have been heralded as strong evidence of extraordinary phenomena is listed in Table 2-1. A number of the explanations are conceded by E. J. Ruppelt and by D. R. Saunders and R. R. Harkins, who are widely regarded as "enthusiasts."

The table illustrates the difficulty with the so-called residuum of classic, unexplained cases: the "residual cases" keep changing. As old cases are explained, new cases are added, keeping the residuum of unexplained cases more or less constant. One prominent student of UFO's who has called for another scientific investigation, included the Santa Ana, Fort Belvoir, and McMinnville cases as unexplained in 1966–1969; yet he later agreed with the Colorado Project's conclusions that the Santa Ana case was internally inconsistent and the Fort Belvoir case is explained. The Colorado Project found McMinnville was inconclusive, and in late 1969 it was shown to be internally inconsistent by R. Shaeffer (private communication). So I must approach UFO's with what I hope is healthy skepticism.

To discuss UFO's in a scientific way we must define terms. For "UFO" I accept the definition of the Colorado Report: essentially it as an unidentified object or apparition considered strange by the observer. This is popular usage, but it leaves a need for other terms to make the discussion clear. A UFO can become an *identified flying object* after investigation. If not, it may be considered an *extraordinary flying object,* that is, something beyond the bounds of recognized natural phenomena. Even more startling, it might be called an *alien flying object* meaning a vehicle constructed by alien intelligence. I therefore speak of an alphabetic spectrum: UFO, IFO, EFO, AFO.

*Table 2-1.* The fate of some of the "best" UFO reports

| Date | Case name | Alleged events | Current status |
|---|---|---|---|
| 1948 Jan. 7 | Godman AFB, Ky. | Pilot Mantell chased UFO, got too close, and was "shot down." Plane crashed and Mantell was killed. | UFO was a skyhook balloon, little known at that time. Mantell flew too high. (MB, SH) |
| 1948 July 24 | Eastern Airlines (over Georgia) | Pilots Chiles and Whitted reported that a fiery cigar-shaped object with windows passed close to airliner. | Description exactly matches that given by witnesses to re-entry of Zond IV in 1968. (CR) |
| 1948 Oct. 1 | Fargo, N.D. | Pilot Gorman has a dog-fight with a lighted UFO at night. | Navy pilots reproduced such an event with a lighted balloon. R, MB identify the object as a balloon, confused with Jupiter. |

*Table 2-1.* Continued

| Date | Case name | Alleged events | Current status |
|---|---|---|---|
| 1950 Mar. 17 | Farmington, N.M. | Hundreds of UFO's witnessed by many citizens. | Identified as fragments of exploded skyhook balloon by both R and MB. |
| 1950 May 11 | McMinnville, Ore. | Farmer and wife photograph disk. | Fabrication not ruled out in CR. Strong internal inconsistencies recently shown. (S) |
| 1952 July 2 | Tremonton, Utah | Navy photographer gets telephoto movies of UFO's. | Identified as birds; similar behavior of birds observed by author near Tremonton in 1968. (CR) |
| 1957 Sept. | Ubatuba, Brazil | Composition of fragments of alleged exploding saucer rule out terrestrial origin. | Composition could have been produced commercially, as established by Dow samples available at the time. (CR) |
| 1957 Sept. | Fort Belvoir, Va. | "Amazing ring-shaped UFO" photographed by Army private. | Product of demonstration explosion: detached vortex ring. Well-shown examples filmed in Soviet production of "War and Peace." (CR) |
| 1963 Dec. 5 | Vandenberg AFB, Calif. | Rocket tracking cameras photograph UFO. | UFO shown to be Venus. (CR) |
| 1965 Aug. 3 | Santa Ana, Calif. | Traffic investigator photographs disk. | Many internal inconsistencies. (CR) |

Sources for current status:

MB: D. H. Menzel and L. G. Boyd, *The World of Flying Saucers* (Garden City: Doubleday, 1963).

SH: D. R. Saunders and R. R. Hawkins, *UFO's? Yes!* (New York: Signet Books, 1968).

CR: E. U. Condon, *Scientific Study of Unidentified Flying Objects* (New York: Bantam Books, 1969) (the Colorado Report).

R: E. J. Ruppelt, *Report on the Unidentified Flying Objects* (New York: Ace Books, 1956).

S: R. Schaeffer, private communication, 1969.

This terminology makes it easier to answer questions such as: "Have there really been any UFO's?" The answer of course is yes. There have been thousands and thousands of UFO's. Many people have reported things they could not understand.

What people usually mean to ask is "Were there any EFO's?" We have to admit, if we are realistic, that there certainly are a few cases that have remained puzzling after analysis. Yet this, too, is to be expected. The tremendous variety of earthly and human phenomena guarantees unexplained reports.

The crux of the argument, a point that UFO enthusiasts tend to overlook, is the variety of ordinary circumstances that can produce apparently extraordinary phenomena. Some of them are listed in the Colorado Report: sundogs, sub-suns, balloons, lenticular clouds, industrial detergents, satellite re-entries, meteor radar echoes, radar angels, and so on. No ordinary observer is expected to be familiar with them all, and no UFO investigator can anticipate all of them and all their combinations.

Therefore we must expect UFO reports whether there are extraordinary flying objects or not; unexplained observations are not evidence for EFO'S. My colleague, the late James McDonald, has asked: "If UFO's really exist, why don't pilots see them? Why don't astronomers see them?" He then showed that pilots and astronomers had reported UFO's. But this seems to imply that UFO's are a single class of objects. The UFO's reported by various pilots, astronomers, and city dwellers may all be results of different individual circumstances. McDonald states: "One can be fooled, of course; but it would be rash indeed to suggest that the thousands of UFO cases now on record are simply a testimony to confabulation." This is the "where-there's-smoke-there's-fire" theory. Similarly, *The UFO Evidence,* a comprehensive study published in 1964 by NICAP (National Investigations Committee on Aerial Phenomena), lists tabulations of pilot-witness, scientist-witness, radar, photographic, and other kinds of cases, and asserts that there must be something to such an abundance of cases. But, of course, the mere listing of unanswered puzzles is not equivalent to providing unanswerable arguments.

Is it conceivable that all of the UFO reports can be due to mistakes and hoaxes? I think that it is conceivable, and not at all a rash suggestion. We know the "signal-to-noise ratio" in UFO studies is low, and there may be no signal at all.

A few examples will illustrate how mistakes and hoaxes can produce UFO reports, or EFO reports, and in particular that secondhand accounts cannot be trusted. A friend brought me what appeared to be two rather crude fake UFO photos and asked what I thought of them. When I told him, he said, "They couldn't possibly be fake because the fellow who took them is a friend who is a sociology professor with outstanding credentials as a witness." A few days later my friend returned sheepishly to tell me that the sociologist showed fake pictures to his friends because he was interested in studying their reactions.

As another example, during the Colorado Project we heard many stories of UFO films hidden away on military bases. The only documented case was cited in a text on astrodynamics.[2] A peculiar image had been photographed passing a rocket during a launch at Vandenberg AFB. Triangulation indicated that the object was distant. We requested the films and classified tracking data through Air Force channels, received them at once, and found that the telephoto trackers had recorded the planet Venus. These photos had stimulated many secondhand rumors that the Air Force had "proof of flying saucers." In 1968, during the hearing on UFO's, of the House Committee on Science and Astronautics,[3] misleading testimony was given that the Colorado Project had been unsuccessful in obtaining or explaining the Vandenberg films. This is an example of how the Congress, the scientist, and the concerned citizen gets incorrect information on UFO's. Lack of scientific communication has allowed a distorted UFO mythology to develop.

Such examples do not prove that the excitement over UFO's is baseless. If we assert that there are no extraordinary objects, we must know exactly how the UFO affair developed. I will review UFO history in the next pages. First, we divide the UFO phenomenon into two parts: the sociological UFO phenomenon and the (hypothetical) physical UFO phenomenon, i.e., the possibility of extraordinary physical objects. The following is a hypothetical reconstruction of how the sociological UFO phenomenon could have developed without any EFO's.

Kenneth Arnold reported his "flying saucers" in June 1947, a postwar era when people were conscious of the possibility of interplanetary travel and the technological marvels that lay ahead. Arnold's sighting was an honest misinterpretation of some ordinary phenomenon, possibly a group of aircraft. At that time, the American public was "primed" to seize upon flying saucers with enthusiasm and excitement. The Maury

Island hoax of July 1947, where alleged fragments of a UFO were first reported, and other hoaxes prove that some citizens were sufficiently motivated to go to great lengths to claim sightings of the strange new objects.

That a society shold be "primed" is no novel concept. The anthropologist A. L. Kroeber pointed out in 1917 that the acceptance of an idea depends as much on the state of society as on the idea itself. Kroeber points to Mendel's discovery of the principles of heredity in 1865. Mendel's work was ignored at the time, but in the year 1900, when Darwin's word had taken hold and the question of hereditary mechanisms was on everyone's mind, three students independently rediscovered Mendel's conclusions with experiments of their own.

In the same way, the sociological reaction to Arnold's report in 1947 can be explained by the fact that postwar society was primed for the acceptance of alien spaceships. Hence the headlong rush to report saucers, sometimes through hoaxes. The spectacular hoaxes acted as a positive feedback, keeping the subject alive and guaranteeing that newspapers would devote space to legitimate reports called in by honest citizens who spotted things they could not identify.

This surge of publicity caused genuine concern among military and other responsible officials. During a military investigation, Arnold came away with the mistaken impression that the military already knew what the objects were and had good photographs of them. By mid-summer of 1947, he publicized this erroneous impression in magazine articles which gained wide attention. This fostered the "government conspiracy" myth. The fact that two military officers were killed in an airplane crash while investigating the Maury Island incident kindled further rumors. Military concern escalated in 1948–1949 because of three spectacular sightings and the Green Fireball episode, publicized in *Life* magazine. These incidents were billed in the press as (1) a successful attack by a UFO on an aircraft, (2) a flaming cigar-shaped object that buzzed an airliner, (3) a dogfight between an aircraft and a UFO, and (4) a rash of unprecedented bolides. Much later they were accounted for as (1) the airplane's encounter with a skyhook balloon, (2) a probable fireball sighting, matching contemporary descriptions of the Soviet Zond IV spacecraft re-entry, (3) a probable dogfight with a weather balloon, matching a similar incident deliberately staged by Navy pilots, and (4) a series of meteoric objects, possibly of lunar origin. The explanations came late and unheralded, as explanations often do. (See Table 2-1.)

Within the Air Force investigating team, a group of officers now began to believe the UFO's were extraterrestrial. Since this was never accepted by their superiors, rumors were generated in the early 1950's that the Air Force was suppressing the "known fact" that saucers were interplanetary vehicles. In 1952, the famous Tremonton, Utah, motion pictures of UFO's were made by a Navy officer, and submitted to the military. Air Force and Navy analysts who already were prepared to believe in alien space ships concluded that the films showed self-luminous objects, which, if ten miles away, were moving faster than 1,000 miles per hour. The Robertson committee, a group of scientists called to advise the Air Force, examined many cases in the Bluebook files, including Tremonton films, and discovered that the assumption of *self-luminous* objects was unwarranted, that the films probably showed birds and were not good evidence of EFO's or AFO's. After reviewing this and other cases, the Robertson committee concluded UFO's posed no threat to the nation.

Unfortunately, the Air Force decided that the UFO affair should be hushed up so as not to clog security networks with sightings generated by an excited public. Though this move to secrecy reflected the uneasiness of the times, it was a crucial error because it encouraged, instead of discouraging, rumor. By comparing the Tremonton films, the Air Force and Navy reports, and NICAP documents, we can now see how false, but justifiable, rumors spread through UFO circles that the Air Force had "evidence" of EFO's. Certain prominent UFO advocates on the fringes of the military picked up these rumors and alleged that the Air Force had films of multithousand-mile-per-hour "machines" of unknown origin. None of the studies of the Tremonton films were released at the time. In 1968, I observed birds of similar appearance near Tremonton, and the Colorado Project concluded that the Tremonton films showed birds.

If, instead of a policy of secrecy, the Air Force had released the UFO data and encouraged all scientists to come and look at its files in 1952, it is probable that the UFO mystery would have been clarified after a few months of scientific and public excitement and, indeed, healthy curiosity.

A 1951 movie, *The Day the Earth Stood Still,* depicted a flying saucer tracked by radar as it approached and landed in Washington, D.C. Some months later, radar operators at the Washington National Airport convinced themselves that precisely this was happening. On two hectic nights in 1952, on the basis of radar returns and alleged visual sightings, they scrambled jet fighters to search for the invaders. The chases netted no saucers but the resulting nationwide headlines kept the UFO's high in the public mind. The case was carefully studied by concerned officials including the Robertson panel, but public scientific scrutiny was never invited.

An interregnum ensued. While saucers slipped from prominence in the newspapers, a lower stratum of literature, a sort of saucer underground of sensationalizing tabloids and occult magazines, kept the UFO mythology very much alive among devotees who ranged from sincerely interested but poorly informed readers, through fanatics, to a few mentally deranged individuals. Few reporters of UFO stories made a critical effort to research their material—they couldn't afford to; the stories might have disappeared!

Charges of obfuscation by the Air Force led to the formation of a number of civilian investigative groups, the most prominent of which was NICAP, formed in 1956 by individuals who were convinced that the Air Force was suppressing relevant data. The poor organization and performance of the Air Force investigative project did little to refute such charges.

In late November 1957, within a month after the launching of the first two artificial satellites, the monthly UFO report rate shot up by a factor of seven, illustrating the "space-flight effect," by which increased public awareness of space activities results in reports of unidentifieds. Saucer enthusiasts were quick to point to the new sightings as evidence of continued UFO activity. Soon after this initial shock wore off, the number of reports declined, since space activities became more commonplace and the nation's interests were taken up with a new decade and a new president.

The second interregnum was broken only by a second spectacular space accomplishment in July 1964, the first spacecraft photography of Mars. Mars had long appealed to the public as a planet possibly inhabited by "humanoids." The photographs of Martian craters on the front page of every newspaper were enough to trigger the space-flight effect once again. This rash of sightings was followed by the northeast power blackout in November 1965, which some tabloids attributed to UFO's.

The fortuitous re-entry of the Zond IV spacecraft (see Chapter 6, Appendix 3) over the United States on March 3, 1968, led to the discovery of two more UFO effects: the "airship effect," in which observers conceive of moving lights in a dark sky as connected in a single entity, and the "excitedness effect," in which observers with the poorest observations are most likely to submit reports. The two most detailed accounts described a cigar-shaped ship with windows, and one witness said it was so close she could have seen people through the windows, had there been people inside! These effects were detailed in the Colorado Report. They made it clear how thousands of reports could be made and how newspaper editors had a continual stream of UFO reports to play up or down, depending on public interest.

This condensed history of UFO's contains a number of documented examples of mistakes, hoaxes, preconceptions, rumors, and misinterpretations. All help to explain how the sociological UFO phenomenon occurred. But why did it produce such a marked human response? The sociology of UFO's can tell us something about human nature. Jules Verne said, "What one man can imagine, another man can do." I propose a corollary: "What one man can fantasize, another man will believe."

This can explain much of the UFO mythology, and is corroborated by new data. When Arnold reported flying saucers in 1947, an excited populace reported exactly the same thing; yet we can prove the 80 to 100 per cent of their sightings were mistakes and hoaxes. When science-fiction stories and films showed the nation's capital approached by saucers in 1952, radar observers there promptly interpreted an episode of unidentified "angels" as the same thing. When organizations were formed claiming that saucers have landed, or that the Air Force was hiding films proving UFO's are spaceships, members could be recruited. The number of believers depends on the credibility of the fantasy, which depends on the public mood at the time. It doesn't matter whether the fantasy is true or not; in the dogmas of astrology and witchcraft, the lack of empirical evidence seems to have little effect. No one asks the important questions: "Does it work?" "What is the empirical evidence?" "Is it based on more than hearsay?" "Does the theory predict any effect I can observe for myself?"

There are spirits in our subconscious that make us believe. The spirit of paranoia prods, "Believe it because all those scientists are picking on it." The spirit of hypocrisy urges, "Pretend to believe it because it can gain you attention and wealth." The spirit of conservatism says, "Believe it because it fits with the other things you believe" (or more often, "don't believe it because it doesn't fit with the other things you believe"). The evangelical spirit whispers, "Believe it because you want something to defend."

The UFO affair, then, may teach us something about the sources of our beliefs. As *New York Times* science writer Walter Sullivan pointed out in his commentary on the Colorado Report: "One wonders to what extent [conditioning] affects such basic attitudes as our nationalism, our theological point of view, and our moral standards. Are they really founded on logic and the ultimate truth?" [4]

Was there any signal buried in the noise? Were there any extraordinary flying objects? Scientists trying to explain some UFO cases remained perplexed. A curious result emerges if we look at these unexplained cases as a group; they diverge quite radically from the popular concept of a "flying saucer." Among them there are no reports of disk-shaped metal ships, no landing gear, and no evidence of intelligence save for the Socorro case, where a vehicle leaving a curious array of imprints in soil was reported near White Sands Proving Ground. Instead, the image created by the puzzling cases is of amorphous glowing objects with dimensions of feet or yards, sometimes recorded by radar and sometimes seen visually, with occasional consistent testimony of automobile electrical system failures such as in the 1957 Levelland, Texas, case, where many independent witnesses reported such glowing masses.

This is as much as one should say by looking at the puzzling cases as a group; there may be fewer than a dozen that involve phenomena marginally outside the borders of accepted science. This is not a strong statement; and we cannot ask for a multimillion dollar study on the strength of such vague evidence. It will do no good for EFO advocates to list the unsolved cases. Most of the "best" cases suggested by enthusiasts in the past turned out to be solvable. In the statistical sample, most of the UFO data are unreliable, so statistical studies will reveal not EFO's but our conceptions of EFO's.

If the proponents of extraordinary phenomena want to be taken seriously, they must pick *one* case which they agree is strong evidence and invite other scientists to investigate it. We should put the burden of proof on the proponents of EFO's. The only way to convince the scientific community that something strange is going on is to present specific evidence concisely. If the evidence is good, the case will stand up, and the existence of extraordinary phenomena will have to be taken seriously.

*Addendum, May, 1972:* I proposed that with publication of the Colorado Report the burden of proof had shifted from skeptics to UFO enthusiasts. Condon, in the Colorado Report, concluded that little of scientific value would result from further study of available UFO reports and it appears to date that this has been borne out. At the same time, Condon noted that valid proposals for studies of the UFO phenomenon should be seriously considered, and I argued that UFO advocates should find and call attention in the technical literature to a *single* case so strong that it would stand up to scientific scrutiny and be established as a real, extraordinary event. A response in this direction is that the AIAA (American Institute of Aeronautics and Astronautics) journal has recently published detailed accounts of certain UFO reports alleged to be particularly strong. It will be of interest to follow future scientific assessment of these reports.

*Note added in proof:* In view of growing popularity of television science fiction serials, and soon-to-be-published evidence from Mariner 9 that Mars was more clement in the past, one might anticipate a resurgence of UFO interest by the date of this book's publication.

## NOTES

1. E. U. Condon, *Scientific Study of Unidentified Flying Objects* (New York: Bantam Books, 1969).

2. R. M. L. Baker, Jr., and Maud Makemson, *An Introduction to Astrodynamics* (New York: Academic Press, 1967).

3. *Symposium on Unidentified Flying Objects* (the Roush Report), Hearings before the Committee on Science and Astronautics, U.S. House of Representatives, 90th Congress, 2d Session, July 29, 1968 (Washington, D.C.: U.S. Government Printing Office, 1968).

4. Walter Sullivan, Introduction to Condon, note 1, p. xiii.

# 3

## Astronomers' Views on UFO's

FRANKLIN ROACH

The UFO phenomenon involves astronomers since the unidentified objects are often seen in the sky, a domain that astronomers have long considered their own. The speculative interpretation of some UFO's as visitations by nonterrestrial intelligent beings usually causes dismay on the part of astronomers because they appreciate the enormous distances involved in the transport between ports of call. However, the public interest in UFO's has lured university students into the study of astronomy (see Chapter 1). I believe that interest in the speculation about possible nonterrestrial origin of UFO's has stimulated a healthy public probing of the fascinating question of extraterrestrial intelligent life.

In 1961, a multidiscipline group of eleven scientists (referred to as "The Order of the Dolphin") met to discuss the quantitative evaluation of extraterrestrial life, and suggested the formula in Table 3-1 for the

*Table 3-1.* The "Dolphin" Formula

$$N = \frac{1}{10} L$$

N = Number of civilizations capable of communicating with ours
L = Lifetime (in years) of each civilization
d = Distance to nearest civilization (in light-years) calculated from N and the known space density of stars in the vicinity of the sun.

| Values of | Pessimistic | Reasonable | Optimistic |
|---|---|---|---|
| L | 10 | $10^4$ | $10^7$ |
| N | 1 | $10^3$ | $10^6$ |
| d | "We *are* alone" | 1400 | 140 |

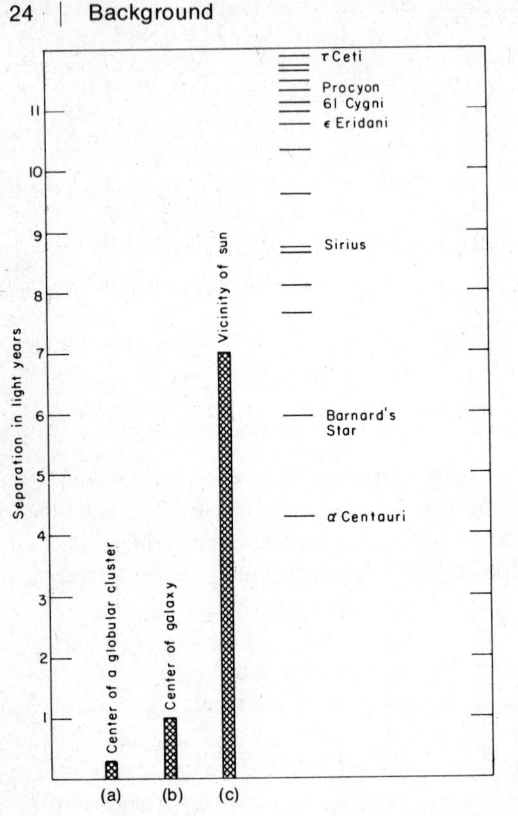


*Figure 3-1.* Average separation of stars one from another, with distances from the sun to nearby stars.

number of civilizations that could be expected to be making an effort to signal some form of information to other neighboring civilizations.[1] Efforts to detect such signaling with large radio receivers (Project OZMA at the National Radio Astronomy Observatory) from two relatively nearby stars were not successful. The technology of such communication involving the speed of light (or of radio waves) is far easier than the technology of physical travel involving speeds less than the speed of light, and I believe most astronomers would agree that the analyses of William Markowitz and of Edward Purcell preclude the possibility of interstellar travel [2] (but see Carl Sagan's discussion in Chapter 14).

The distances between stars are conveniently expressed in light-years —a distance unit which has the advantage of carrying with it a time implication. In Figure 3-1 some measured stellar distances are shown, and the mean separations are shown graphically (a) for stars in the solar neighborhood, (b) for stars near the center of our galaxy, and (c) for stars in the center of a rich globular cluster. Note that interstellar travel involves distances of several light-years for travelers in our part of the galaxy, about one light-year near the center of the galaxy and one or two light-months between stars in the center of globular clusters. We may speculate that travel skills have been perfected by globular-cluster cosmonauts, but this does not help us in the UFO problem since the globular clusters are thousands of light-years from our solar system.

Harlow Shapley [3] has emphasized that interstellar space must include a large number of objects having masses somewhere between that of our giant planet Jupiter (1 / 1000 that of the sun) and a very faint star (say 1 / 100 that of the sun). I have attempted to make a quantitative estimate of the interstellar destiny of these "Lilliputs" in Figure 3-2, a log-log plot of the cumulative number density of objects versus the mass of the individual object. The data for stars are from standard tables of the so-called luminosity function combined with the mass-luminosity re-

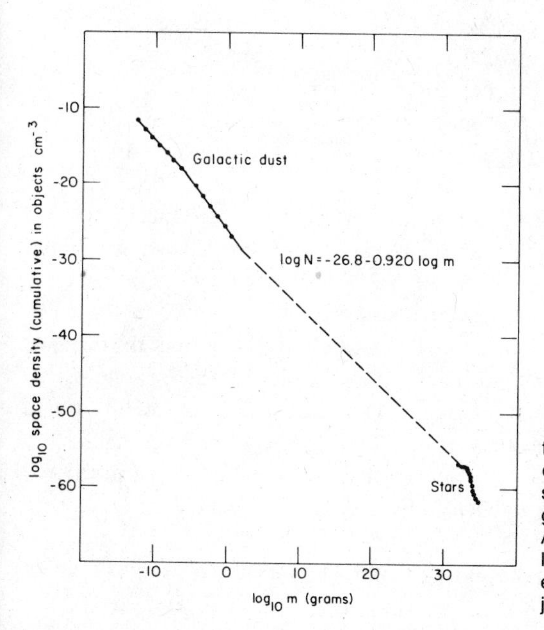


*Figure 3-2.* An attempt to interpolate the space density of objects having sizes between those of galactic dust and stars. Any reasonable interpolation indicates the presences of "Lilliputian" objects.

lationship for stars. At the other end of the mass scale, the amount of galactic dust is based on the measured extinction of starlight by interstellar dust. The interpolated (dashed) curve is used for estimating the space density of objects of intermediate mass. For example, Jupiter has a mass of $1.9 \times 10^{30}$ grams, and the interpolated curve gives for the space density of Jupiter-like objects, $2.2 \times 10^{-55}$ per cubic centimeter, or 0.19 per cubic light-year. The mean separation of these objects in space is 1.75 light-years and the closest one to an average observer is 0.97 light-year. If this space density holds for the vicinity of the sun, the nearest Jupiter-like object would be about one-quarter as far as the nearby star, Alpha Centauri. Shapley has suggested that Lilliputs somewhat more massive than Jupiter should have an internal source of heat and might spawn life. Therefore, the nearest life outside our solar system might be found on such Lilliputs rather than on planets in orbits around other stars. If this life is "intelligent" could "they" be signaling or visiting us?

In the vast domain within a sphere halfway to Alpha Centauri there should be ten objects having a mass of Jupiter, 1,000 of Earth size, 60,000 of moon size, and five million of the size of the asteroid Ceres (mass $7.5 \times 10^{22}$ grams). The prediction for comets (assumed mass $10^{18}$ grams) comes out to be $10^{12}$. Some time ago Jan Oort [4] suggested that the comets we observe may have been deflected from nearly circular orbits in this vast region between the sun and nearest stars so that they "fall" in toward the sun. In the same way, during billions of years a Lilliput may have been sufficiently perturbed to come much closer to the sun than its original one light-year. If so, this "Planet X" would have the view described in Table 3-2, where our view is given in the last two lines, for three assumed distances. Space travel to the earth from 256 AU (astronomical units; 256A.U. $\approx$ 100,000 times the distance from the earth to the moon) is certainly possible, but there is one speculation on another speculation here: (1) that on a Lilliput intelligent and technologically advanced forms of life will develop, and (2) that such a Lilliput has been perturbed into an orbit close enough to permit travel to Earth and yet large enough to have escaped detection. At least there is plenty of time (some 4 or 5 billion years) for these astronomical developments. For example during 4 billion years there would be about

*Table 3-2.* Planet X (Mass: 5×Jupiter; Diameter: 1.7×Jupiter)

| Feature | Unit | Close | Medium | Far |
|---|---|---|---|---|
| Distance from sun | A.U. | 256.0 | 2560 | 25,600 |
| | Light-days | 1.5 | 15 | 150 |
| | Miles | $7.4 \times 10^{10}$ | $2.4 \times 10^{11}$ | $2.4 \times 10^{12}$ |
| From X | | | | |
| Brightness of sun | m(vis) | −14.5 | −9.5 | −4.5 |
| Brightness of Jupiter | m(vis) | 6.4 | 11.4 | 16.4 |
| Separation (Jupiter and sun) | Arcminutes | 70 | 7 | 0.7 |
| Angular diameter of sun | Arcseconds | 7.5 | 0.75 | 0.075 |
| Annual movement of sun | Arcseconds | 137 | 10.2 | 0.32 |
| From Earth | | | | |
| Brightness of X | m(vis) | 13 | 23 | 33 |
| Period of revolution | years | 4100 | 130,000 | 4,000,000 |

30,000 circuits of the sun by a Lilliput having an aphelion of 60,000 AU and a perihelion of 2,500 AU.

The lesson to be learned from this astronomical time scale is that we must be cautious in placing technological limits on other civilizations which are older than ours by some period like a megayear (one million years). "Megaphysics" may even transcend Markowitz' metaphysics.

What are the views of astronomers on the subject of UFO's? There is no single statement that represents any degree of unanimity among astronomers. The physical and temporal nature of our part of the Milky Way galaxy is generally accepted, since it is based on careful measurements repeated and cross-checked, but the significance of UFO's will be debated. One view is that all could be readily explained as "natural" phenomena if we had better data. The other extreme leads to the hypothesis of extraterrestrial visitations.

I remember a social gathering early in the Colorado Project, which I worked on, when Dr. Hynek summed up his attitude with the words "I am puzzled." My reaction was "I am curious," and this led me into the UFO study. Later my curiosity developed into an interest in the general problem of extraterrestrial life (quite apart from possible visitations) and I was especially intrigued by the late Otto Struve's interest in the problem which developed from his work on the rotation of stars.

Many astronomers would agree with Gerard Kuiper, who said:

> I should correct a statement that has been made that scientists have shied away from UFO reports for fear of ridicule. . . . A scientist chooses his field of inquiry because he believes it holds real promise. If later his choice proves wrong, he will feel very badly and try to sharpen his criteria before he sets out again. Thus, if society finds that most scientists have not been attracted to the UFO problem, the explanation must be that they have not been impressed with the UFO reports.[5]

Astronomers view the sky frequently, and it is to be expected that UFO's or "UBO's" (unidentified bright objects) would be observed incidentally in the course of their systematic observation by optical or photometric techniques. Thornton Page has made an interesting analysis which summarizes the fractional sky coverage by the battery of some 300 active professional astronomical telescopes in regular use on the surface of the earth (see Table 3-3). He finds the interesting result that a 30-degree cone of sky centered on the zenith has a coverage of only 1.5 per cent (see the last entry in column 6 of Table 3-3).

*Table 3-3.* Astronomical telescopes of the world

| Class of telescope | Number n | Solid angle of observed field (square degrees) | No. of photos per year | Exposure t (hr) | Sky coverage (nC *) |
|---|---|---|---|---|---|
| Large Schmidts | 3 | 36.0 | 3000 | 0.400 | 0.00518 |
| Medium Schmidts | 27 | 20.0 | 3000 | 0.100 | 0.00649 |
| Small Schmidts | 17 | 15.0 | 2000 | 0.100 | 0.00204 |
| Smithsonian Network | 12 | 100.0 | 5000 | 0.002 | 0.00048 |
| Astrometric | 39 | 1.0 | 1000 | 0.100 | 0.00016 |
| Other (photo only) | 211 | 0.1 | 500 | 0.500 | 0.00021 |
| Total coverage | | | | | 0.01456 |

Source: Adapted from Thornton Page, "Photographic Sky Coverage for the Detection of UFO's," *Science* 160(1968):1258. Copyright 1968 by the American Association for the Advancement of Science.

* nC = fractional coverage of a 30-degree cone centered on the zenith by *n* telescopes.

The following list shows other types of photographic and photoelectric sky coverage, and the scientists investigating them.

Astronomical photography (Page)
Meteor networks
  Prairie Network (U.S.A.) (Page, Ayer)
  Canadian Network
  Czechoslovak Network
The Tombaugh Search (Tombaugh)
  For terrestrial "moons"
  For trans-Neptunian planets
All-sky auroral cameras (Ayer, Rothberg)
Airglow scanning photometers (Ayer, Roach)

Of particular interest is the systematic survey made by Clyde Tombaugh (as yet unpublished) in searches for terrestrial moons and trans-Neptunian planets. It is probable that he would have found any Lilliput as close as 250 astronomical units. Table 3-4 shows a comparison of the Smithsonian Astrophysical Observatory's Prairie Network for the detection of bright meteors and the airglow photometer network used during the International Geophysical Year (July 1957–December 1958). These records should be examined for UBO's.

*Table 3-4.* Comparison of Prairie Network and airglow photometers

| Feature | Prairie Network | Airglow photometers (I.G.Y.) |
|---|---|---|
| Number of stations | 16 | 25 |
| Number of instruments | 64 | 25 |
| Area of Earth's surface "covered" | 1,500,000 km$^2$ | 454,000 km$^2$* |
| Fraction of Earth's surface "covered," $F$ | 0.0029 | 0.0004 |
| Limiting magnitude (fixed) | +4 | 3 |
| Limiting magnitude (moving) | −4 | 3 |
| Fraction of time operating, $f$ | 0.23 | 0.1 |
| Total coverage (fractional) | 0.00067 | 0.00004 |
| D, reciprocal | 1500 | 250,000 |

* Assuming a depth of detection of 10 km

My own research for more than three decades has been involved with the study of the light of the night sky and of the airglow. In general, the records are not examined for UBO's. Although it can be stated that over a considerable period of time no UBO's were reported, it definitely *cannot* be stated that UBO's were absent from the records. In the study of the photometric records the stars are underdrawn (their light omitted as in Figure 3-3), and any UBO which gave a starlike deflection would be missed. During the Colorado Project, Frederick Ayer [6] supervised the detailed study of one night of observations at Haleakala Observatory in Hawaii in which the analysts were instructed *not* to underdraw any deflections at all. All starlike deflections were then compared with the positions of known stars and planets. Somewhat to our surprise, on

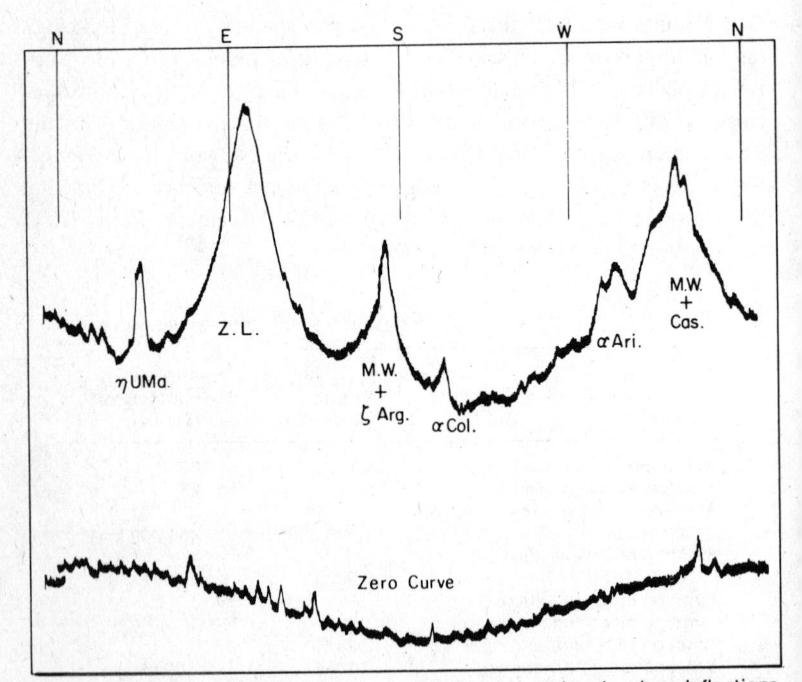


*Figure 3-3.* Typical photometric record around the sky showing deflections due to individual stars (*eta* Ursae Majoris, *alpha* Columbae and *alpha* Arietis) to the Milky Way (M.W.), and to the zodiacal light (Z.L.)

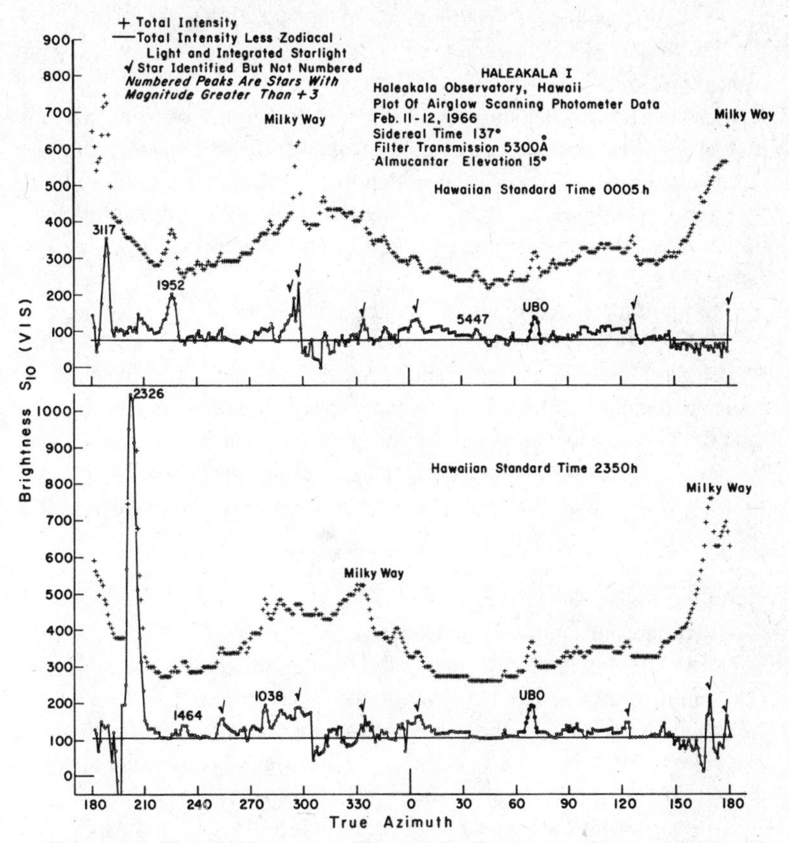


*Figure 3-4.* Photometric record like that in Figure 3-3 showing signals labeled UBO not due to known astronomical objects. Adapted from Condon, *Scientific Study of Unidentified Flying Objects.*

two of the records near midnight there were unmistakable deflections *not* due to known astronomical objects (Figure 3-4). An attempt to identify the objects responsible for the deflections as a suborbital missile crossing the Pacific from the west coast (Vandenberg Air Force Base) to the Kwajalein area was unsuccessful.

The lessons to be learned from this are: (1) it is important to distinguish carefully between the lack of reports and the lack of systematic search, and (2) although it is possible to say something about what the UBO's on the records are *not,* this does not provide information on what they *are.*

J. Allen Hynek [7] has introduced the useful idea of a two-dimensional plot of UFO reports (see Figure 4-1) from which we can judge the potential value of an investigation in depth. His coordinates are "credibility" and "strangeness": if both are high, the sighting should be followed up; if either is low, energy and time should be conserved by filing the report away.

The photometric readings mentioned above should rate very high in credibility, but it is difficult to know how to evaluate the "strangeness." Of course it is strange that we can't identify the deflections, but the fascinating details that often accompany visual sightings cannot be obtained from an impersonal galvanometer record, which tells us only that the unidentified object was as bright as a second-magnitude star. (This suggests a correlation between "strangeness" and the personality of the reporter.)

## Conclusions

My conclusions must be considered as entirely independent of my rather small contribution to the Colorado Report, although I am naturally influenced by my participation in the investigation.

I think it likely that there are many extraterrestrial civilizations in our galaxy, but I think the evidence of UFO sightings does *not* support the hypothesis of visitations by these extraterrestrials.

Direct exploration of the planet Mars before the end of this century should answer the question of life—simple or complex, now or in the astronomical past—on that planet and indicate the relationship between the development of life forms and the physical environment.

Interstellar travel poses problems not soluble with our present technology, or even with developments over the next century. I leave open, however, the nature of the technology based on what I have called "megaphysics." Travel between the earth and a hypothetical Lilliput in an eccentric orbit is an intriguing idea. I recommend that astronomers during the next megennium keep their eyes open and strive, along with all our terrestrial companions, to make the quantity $L$ in the Dolphin formula even greater than a megennium.

## NOTES

1. A general treatment of the "Dolphin Formula" may be found in I. S. Shklovskii and C. Sagan, *Intelligent Life in the Universe* (San Francisco: Holden-Day, 1966), Chap. 29.

2. William Markowitz, "The Physics and Metaphysics of UFO's," *Science* 157 (1967): 1274; Edward Purcell, "Radio Astronomy and Communication through Space," reprinted in A. G. W. Cameron's book on *Interstellar Communications: Collection of Reprints and Original Contributions* (New York: Benjamin, 1963).

3. Harlow Shapley, "Crusted Stars and Self-Heating Planets," *Matematica y Fisica Teorjca* 14 (1962): 69; see also p. 60 of Shapley's book *Of Stars and Men* (Boston: Beacon Press, 1958).

4. Jan Oort, "Empirical Data on the Origin of Comets," from *The Moon, Meteorites and Comets,* ed. B. Middlehurst and G. P. Kuiper (Chicago: University of Chicago Press, 1963).

5. Gerard Kuiper, presentation to the Arizona Academy of Science Meeting, April 29, 1967; reprinted as Appendix C of the Colorado Report.

6. Section VI, Chap. 9, of the Colorado Report.

7. See, for example, his review of "The Condon Report and UFO's," *Bulletin of the Atomic Scientists,* 25 (1969): 39.

# PART II

## OBSERVATIONS

# 4

## Twenty-one Years of UFO Reports

J. ALLEN HYNEK

My role here today is that of reporter; to report to you on my score or so years of experience with UFO *reports* (note that I do *not* say UFO's, for I myself have never had a UFO experience) and with those who make such reports, from this and many other countries. I was asked in 1948, as an astronomer then at Ohio State University, and thus geographically near the Wright-Patterson AFB, to review the UFO reports received by the Air Force and to determine how many of them originated from misperceptions of astronomical objects or events. This consulting role continued across the years and gave me the chance to monitor the flow of UFO reports submitted to the Air Force, and to observe the Air Force handling of the problem as first one, then another officer took charge of Project Bluebook.

As reporter of the UFO scene, I am reminded of the old dictum of the reporter: find out Who, What, Where, When, and Why. I will have no difficulty in dealing with the Who, What, Where, and When, for that means simply dealing with facts—particularly with the incontrovertible fact that UFO *reports* exist, and that the time and location of the reported event is generally known, as well as the identity of the witnesses.

The "Why" I shall leave to other scientists, but I shall challenge their explanations if they are not conversant with the Who, What, Where, and When. I am very weary of pontifications by those who have not done field- or home-work, so to speak.

Indeed, I would like to say a word about scientific methodology as it pertains to this problem. I have discussed this at length with the noted Canadian philosopher of science, Thomas Goudge.

"One of the most interesting facets of the UFO question to me," Goudge writes, "is its bearing on the problem of how science advances. Roughly I would say that a necessary condition of scientific advance is that allowance must be made for (a) genuinely new empirical observations and (b) new explanation schemes, including new basic concepts and new laws." Goudge notes that throughout history any successful explanation scheme, including twentieth-century physics, acts somewhat like an "establishment" and tends to resist genuinely new empirical observations, particularly when they have not been generated within the accepted framework of that scheme— as, for instance, the reluctance to accept meteorites, fossils, the circulation of the blood, and, in our time, ball lighting. History is replete with such examples. When the establishment does accept such new observations it often tends to assimilate them into the going framework—as, for instance, the attempt to admit the existence of meteorites as stones that had been struck by lightning. "Hence," Goudge concludes, "the present establishment view that UFO phenomena are either not really scientific data at all (or at any rate, not data for physics) or else are nothing but misperceptions of familiar objects, events, etc. To take this approach is surely to reject a necessary condition of scientific advance." [1]

We will never know whether UFO reports represent genuinely new empirical observations if we continue the type of logical fallacy illustrated by the Air Force analysis of a radar-visual UFO report from Kirtland Air Force Base, Albuquerque, New Mexico, in 1957 (see Selected UFO Cases, above). Two witnesses in the control tower reported at 11:00 P.M. that an object, which looked (through binoculars) like a lighted, up-ended automobile, came within 200 feet of the ground when it disappeared behind a fence in a highly restricted area, easily visible from the control tower, then rose abruptly at very high angular rate and disappeared. It was observed visually for about six minutes, about half of that time through binoculars, and tracked in part by radar. The report of the Air Force officer who investigated this case, which is in the Bluebook file, states:

> The two sources are Airways Operations Specialists with a total of 23 years experience. Both were on duty in the control tower at Kirtland Air Force

Base when the sighting was made—both appeared to be mature and well poised individuals, apparently of well above average intelligence, and temperamentally well qualified for the demanding requirements of control tower operators. Although completely cooperative and willing to answer any question, both sources appeared to be slightly embarrassed that they could not identify or offer an explanation of the object which they are unshakably convinced they saw. In the opinion of the interviewer, both sources are completely competent and reliable.

Project Bluebook explained this sighting as that of an aircraft; and gave the following specific reasons:

1. The observers are considered competent and reliable sources and in the opinion of the interviewer actually *saw* an object they could not identify.
2. The object was tracked on a radar scope by a competent operator.
3. The object does not meet identification criteria for any other phenomenon.

So, the witnesses were solid, the radar operator competent, and the object unidentifiable as any other phenomenon; *therefore* the object had to be an aircraft. Clearly, if such reasoning is applied to all UFO reports we can hardly expect to find out whether any genuinely new empirical observations exist to be explained. Schroedinger, the father of quantum mechanics, wrote: "The first requirement of a scientist is that he be curious; he must be capable of being astonished, and eager to find out." Perhaps he should have added, "and be ready to examine data even when presented in a bewildering and confusing form."

There is much in the UFO problem to be astonished about—and much to be confused about, too. Such confusion is understandable. Over the past twenty years I have had so many experiences with crackpots, visionaries, and religious fanatics that I hardly need be reminded of people who espouse the idea of UFO's as visitors from outer space for their own peculiar purposes. You will note that I say "espouse the idea," *not* "make UFO reports." Very rarely do members of the lunatic fringe make UFO reports. There are many reasons for this; primarily it is simply that they are incapable of composing an articulate, factual, and objective report.

In addition to being fully aware of the cultists, and how they muddy the waters even though they don't generate UFO sightings, I am also well aware of the widespread ignorance, on the part of many, of astronomical objects, high-altitude balloons, special air missions, mirages, and special meteorological effects, and of people's willingness to ascribe their views of such things to the presence of something mysterious. These people, in contrast to the crackpots, are far more of a problem because they *do* generate UFO reports which represent a high noise level—so high, in fact, that many who have not looked carefully into the matter feel that all UFO reports stem from such misperceptions. In actual fact it is relatively simple for an experienced investigator to sort out and quickly eliminate virtually all of the misperception cases.

It is a pity that people so often are not well-informed, objective, and accurate reporters; since 1948 I have become only too familiar with UFO reports spawned by Venus, twinkling stars, aircraft, and the like. Some eighteen years before the Condon committee was formed I was already aware that the great majority of UFO reports are nothing more than misperceptions by the uninformed. Of course, these misperceptions must be deleted before any serious study of the UFO question can begin. From this point on, I am speaking only of UFO reports which *remain* unexplained by trained investigators; only then are we truly dealing with something that is *unidentified* by people capable of making an identification. In short, an original UFO report must pass through a "narrow band-pass filter" before it qualifies as worthy material for scientific study, the objective of which is to determine whether any genuinely new empirical observations exist. Only those reports which survive the running of this gauntlet can qualify.

An objection to this approach immediately arises: Aren't we just rejecting everything but the tail-end of the distribution curve of human reactions to visual stimuli? I firmly agreed with this view during my first years of association with the UFO problem, but now I question it. We can take the position that we are dealing with the vagaries of human perception only if we are dealing with a *homogeneous* set of observations. For instance, the distribution curve of fruit size in an apple orchard would have a significant tail at the large-fruit end if measurements of watermelons on the ground were included without noting the structural differences between apples and watermelons.

Let me define the UFO phenomenon, the *existence* of which we wish to determine or deny, as that phenomenon described by reports of visual or instrumental observations of lights or objects in the sky (or near, or on the ground) whose presence, trajectories, and general character are not explainable in *verifiable* physical terms, even after intensive study. The Condon Report furnishes us with many examples.[2]

For years I could not accept the idea that a genuine UFO phenomenon might exist, preferring to hold that it was all a craze based on hoaxes and misperceptions. As my review of UFO reports continued, and as the reports grew in number to be of statistical significance, I became concerned that the whole subject didn't evaporate as one would expect a craze or fad to do. Also, the phenomenon of UFO reports persisted not only in this country but in many areas over the world; if there were some worldwide compulsion to report strange things, why are only these particular types of strange reports preferred from the infinite universe of all possible strange reports?

The degree of "strangeness" is certainly one dimension of a filtered UFO report. The higher the *strangeness index* the more the information aspects of the report defy explanation in ordinary physical terms. Another significant dimension is the probability that the report refers to a *real* event; in short, did the strange thing really happen? And what is the probability that the witnesses described an actual event? This *credibility index* represents a different evaluation, not of the report in this instance, but of the witnesses, and it involves different criteria. These two dimensions can be used as coordinates to plot a point for each UFO report on a useful diagram. The criteria I have used in estimating these coordinates are: For *strangeness:* How many individual items, or information bits, does the report contain which demand explanation, and how difficult is it to explain them, on the assumption that the event occurred? For *credibility:* If there are several witnesses, what is their collective objectivity? How well do they respond to tests of their ability to gauge angular sizes and angular rates of speed? How good is their eyesight? What are their medical histories? What technical training have they had? What is their general reputation in the community? What is their reputation for publicity-seeking, for veracity? What is their occupation and how much responsibility does it involve? No more than quarter-scale credibility is to be assigned to one-witness cases.

If one now plots the strangeness (S) of a report against the credibility (P) of the witnesses—that is, the probability that the event happened more or less as stated—one obtains a diagram which may be called the strangeness-probability diagram. An example of such a diagram constructed for some cases I have personally investigated is shown in Figure 4–1. Plotted points represent only those UFO reports, of course, that have passed through the misperception and hoax filter. Clearly, the most provocative and potentially important UFO reports are those in the upper right-hand region of such a diagram, representing reports that contain many information elements and have a high probability rating.

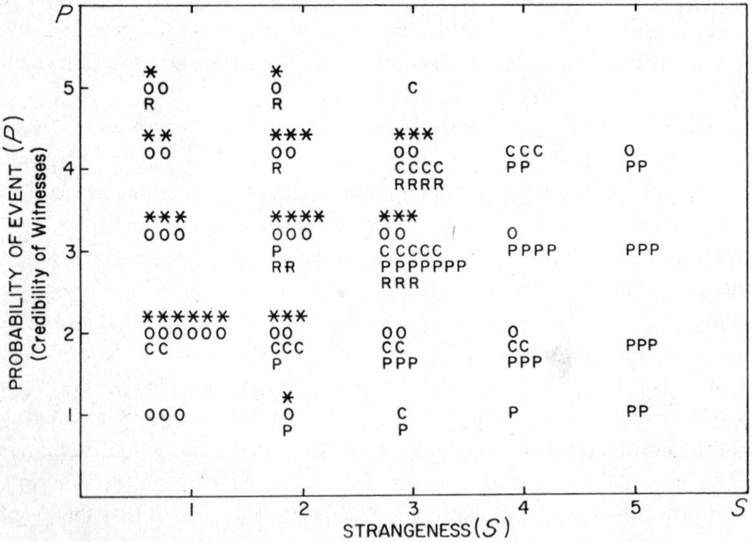

*Figure 4-1.* Strangeness/probability diagram of UFO sightings. To be considered important, such a sighting must have both a very high probability of having actually occurred and a very high strangeness. The upper right-hand corner of the diagram is not heavily populated. * = nocturnal lights; O = daylight disks; R = radar cases; C = close encounters with no interaction with the environment; P = close encounters with physical effects (landing marks, burnt rings, engine stoppages, etc.).

In these high-P reports, the witnesses were of such a caliber and the circumstances surrounding the reported event were such that we cannot discount the reported event. Examples of such information bits are craft description, motions that seemingly defy inertial laws, effects on animals, interference with automobile ignition systems, and visible marks on land. The Condon Report includes several such cases.

My long experience in personal contacts with witnesses who generate high-S high-P reports shows that all were trying to describe an *event* for which they had an entirely inadequate vocabulary—much as an aborigine lacks the vocabulary to describe a supersonic jet or a nuclear submarine. Whatever else can be said of the UFO phenomenon, it represents for the witness an undoubted event, and an event for which he was totally unprepared. The majority of such witnesses, contrary to popular belief, were originally highly skeptical about UFO's. Suddenly they had an experience which affected them profoundly, sometimes traumatically. Faced with the experience of the UFO event, witnesses are generally perplexed and uncertain as to what to do about it. Invariably they attempt to explain it in ordinary terms and fail. Curiosity overwhelms them, yet they know that they will be targets for ridicule if they report (they confess that they had often in the past ridiculed others). Generally they first confide only in their own family, and often prefer to remain silent. Only those who finally report their observations furnish us with data for study.

Any serious investigator is aware that many unreported experiences must exist. Not only has a Gallup poll so indicated, but I frequently try the experiment of asking for a show of hands of those who either have had a puzzling UFO experience themselves or have heard of one from close friends. I generally find that more than 10 per cent of the audience will raise a hand. But when I ask for hands of all those who *reported* the event in some official manner, I find virtually no hands raised. Judging from this and other personal observations, I would estimate that for every officially reported UFO sighting there may exist dozens that have gone unreported.

As scientists we should be astonished that high-S high-P reports have been made in the past five or ten years. What does a serious person with a valued reputation stand to gain by making such a report? Why do people go to the trouble of filling out questionnaires, of subjecting themselves to sometimes hostile inquiry, and of being the target of unpleasant attention?

The reason appears to be twofold. Witnesses have told me that they had not intended to say anything about their experience but they felt that it might be of importance to the government, or to science, and felt it their duty to report. The second reason is curiosity. They want to know whether anyone else experienced the same event, and whether the event has a rational explanation. They are visibly reassured when I tell them, if it be the case, that their sighting fits a pattern and resembles other reported sightings from various parts of the world.

What are the patterns of UFO reports? How can we classify UFO reports (after screening) as an aid to their study? Clearly, if each UFO report represents a unique happening, the UFO is not amenable to scientific study. Such a classification, however, must be free of any preconceived ideas as to the nature and cause of UFO's. Thus the classification must be descriptive; it should be similar to the classification of stellar spectra in the days before we had a theory of stellar spectra, or somewhat like the classification of galaxies today.

I have adopted a very simple classification system based solely on the manner of observation. Such a system tells us nothing, of course, about the nature of the UFO, but it can suggest a means of gathering further data.

There seem to be four basic ways in which the UFO presents itself, so to speak, for human observation: (1) as "nocturnal lights," the objects to which the lights are presumably attached being generally barely, if at all, discernible, (2) as "daylight disks," when the UFO generally, though not necessarily, appears as a disk or long oval; (3) as "close encounters" during day or night: these are sightings made at ranges less than 1,000 feet and often accompanied by physical effects on the land, on plants and animals, and occasionally on humans; and (4) radar UFO's, a special subset of which is the radar-visual observation, in which the radar and visual observations are mutually supporting. These observational classifications are not meant to be mutually exclusive. Clearly a nocturnal light might be a daylight disk in the daytime, and both might become close-encounter or radar cases.

Let us examine each category. A nocturnal-light report offers the least potential for scientific study, as it has the fewest information elements and thus a low strangeness index. The nocturnal-light UFO can be defined as a light or combination of lights whose kinematic behavior passes through the "UFO report filter"; that is, it cannot be logically ascribed to balloons, aircraft, meteors, planets, satellites, satellite reentries, or missiles. The experienced investigator generally has no difficulty with the screening process here. Years of checking enable him to filter these out almost at first glance. Of course, should a UFO choose to masquerade as a hot-air balloon or a photographic night-air exercise, there is no easy way of differentiation, at least so long as we are limited to observing from the ground. If we had immediate reaction capabilities, and could dispatch an interceptor aircraft, then we could clear the matter up quickly, or perhaps we would experience what has often been reported in the past twenty years: as the intercepting plane approaches the light in question, the light either suddenly goes out or seems to take off and soon outdistances the investigator. In that event the report earns its place as a high-S high-P member of the nocturnal-light category.

An example of this category is a case I investigated personally, involving five witnesses, the senior witness being the long-time associate director of a prominent laboratory at MIT. The nocturnal light was first sighted by his son, who had been out airing the dogs. He came bounding into the house crying, "There's a flying saucer outside!" The senior observer picked up a pair of binoculars on his way out. He told me that he didn't expect to see anything unusual but was going out to see what the commotion was all about. For the following ten minutes he was engrossed in what he saw—the nature of the light, its motions, its hovering, and its take-off. He described the light as having a high color-temperature although essentially a point source, subtending less than a minute of arc in the binoculars. The five observers were fortunately able to compare it to an airliner and a helicopter, both of which passed by during the observation interval, and neither the motions nor lights of these craft bore any resemblance to those of the UFO, subclass NL. The trajectory of the object was plotted against the framework of the branches of a denuded tree. This observer was a good one, and his report included the condition of his eyes and those of the members of his family. The adult observers were both far-sighted and the senior observer wore glasses only for reading.

Incidentally, all my attempts, as scientific consultant to the Air Force, to mount a serious investigation of this case came to naught. The Bluebook evaluation is "unidentified," but somehow this word is not a challenge to inquiry. It has been classified as "unidentified," and therefore the case is "solved": it has been identified as "unidentified"!

So certain is the Air Force, at least publicly, that all UFO reports must represent normal things that they see no point to serious investigation. During most of the time I acted as their consultant I repeatedly urged immediate reaction capability and proper scientific investigation, but to no avail.

The second category, the "daylight disk," covers reported daylight sightings of objects seen at moderate distances. The prototype report runs something like this: I was driving along and there crossed over in front of me a shiny metallic disk. It seemed about 500–1,000 feet above the road. It came down fairly close to the ground, stopped and hovered with a wobbling motion and then took off with incredible speed, straight up, and was gone in a few seconds. There was no noise. This category understandably has more photographs to support it than all the others put together. An example is the McMinnville, Oregon, case which the Condon Report lists as unsolved (Case 46).

A photographic daylight disk case was reported by three prospectors in bush country near Calgary, Alberta. I personally investigated the terrain, the people, the negatives, and the camera. Fred Beckman of the University of Chicago and I have satisfied ourselves that the images on these color negatives are real images. The terrain, the interrogations of the witnesses, and the sworn affidavit of the principal witness all lead me to ascribe a high SP rating to this case.

The published literature on UFO's has many photographs. Some are clearly hoaxes, but many have never been investigated sufficiently to rule out very sophisticated hoaxes. A hoax is all one has to rule out, however. For if the daytime photo shows any detail at all, aircraft, balloons, and so forth may be immediately eliminated. The picture itself is sufficient to establish the strangeness index; it is the credibility index that is difficult to assess. Proper interrogation, tracing of the processing history of the negative, microscopic and microphotometric examination of the negative, plus proper psychological testing of the witnesses to the taking of the photograph should serve to rule out all but the most highly sophisticated, expensive, and laboriously contrived hoaxes. In any one case it is clearly impossible ever to state unequivocally that a photo of a daylight disk is genuine, but I would submit that twenty-five such separate photographic cases, each subjected to exhaustive tests, would allow us to say that the *probability* of a hoax in all twenty-five cases is vanishingly small.

Even this does not prove the existence of truly strange flying objects, but it should be sufficient to attract the proper attention of the scientific world. That, of course, has long been my position: that some UFO reports are worthy of serious scientific attention. Inherent in the sheaves of UFO reports may well be many doctoral dissertations for physicists, sociologists, and psychologists alike.

The third category of UFO reports, the "close encounter," offers by far the greatest potential for scientific study. Since a close encounter obviously offers a greater chance for observation, we can expect, and we get, many more information elements, and hence a higher strangeness index. Here the theory of simple misperception fails utterly in explaining reports of craft landing 100 feet away, of visible marks left on the ground, of animals and people visibly affected, and of automobiles temporarily stopped on the road. Either we must say that the witnesses were mentally unbalanced or that something most interesting actually happened. However, I am not taking sides; I am merely reporting to you what is reported all around the world, and by seemingly competent witnesses.

I divide the close encounter cases into three subdivisions: the close encounter, with little detail; the close encounter with physical effects; and the close encounter in which "humanoids" or occupants are reported. This latter subgroup, of course, has the highest strangeness index and frightens away all but the most hardy investigators. I would be neither a good reporter nor a good scientist were I deliberately to reject data. There are now on record some 1,500 reports of close encounters, about half of which involve reported craft occupants. Reports of occupants have been with us for years but there are only a few in the

Air Force files; generally Project Bluebook personnel summarily, and without investigation, consigned such reports to the "psychological" or crackpot category.

A prototype of the simple close encounter goes like this: Witnesses are driving along a lonely road when the driver spies a strange glare in his rear-view mirror. He becomes frightened and increases his speed, trying to outdistance the UFO, but he cannot. He stops the car and tries to take cover. Shortly the light rises and vanishes quickly in the distance. It is easy to say that such witnesses are mentally unbalanced, unless one must say it to their faces, especially knowing that they are respected members of their communities and hold responsible positions.

The close encounter with physical effects is the category that interests me most, since the reported effects on animal, vegetable, and mineral are potentially measurable. For instance, there are more than a hundred reports on record of UFO's that reportedly caused car ignition failures. In a typical case a bright light suddenly appears and seems to seek out the witnesses' car. As it stops to hover over the car, the car lights dim, or fail, and the engine dies. Often the occupants of the car report feeling hot and prickly. After a few minutes the apparition leaves, and the car returns to normal operation, but the witnesses often do not; their equanimity is temporarily destroyed.

Witnesses of such encounters do not readily submit themselves to interrogation. Often they tell no one for days, or they tell only very close associates. Eventually a serious UFO investigator comes to hear about it, and then the story unfolds. If they unwisely tell their story indiscriminately, their lives are made miserable by ridicule and taunts of unsympathetic friends.

Let us consider the probabilities in car-failure cases. On the road we occasionally pass a disabled car, its hood up, waiting for the repairman or the tow truck. We should regard it as odd, and of low probability, if the car were to heal itself after a few minutes and proceed as though nothing had happened. If we add that this event was accompaned by a very bright unexplained light hovering over the car, then I submit that the probabilities are extremely small. And when we deal with dozens of such cases, we are driven to the conclusion that something most extraordinary must have happened. If we have in these cases what Goudge calls "genuinely new empirical observations [requiring] new explanation schemes," then we can anticipate an exceptional scientific breakthrough, although it may not be just around the corner.

In this twentieth century we may be as far from a solution of the UFO problem as nineteenth century physicists were from an interpretation of the aurora borealis. Even so, it is incumbent upon us as scientists to document and study the phenomenon to the best of our ability. But the present lack of continued scientific study still leaves it unclear whether genuinely new empirical observations exist. Even the Condon Report left unexplained about one-quarter of the cases examined.

The fourth observational category contains those UFO reports involving radar. There are many reports in this category from responsible persons, such as pilots and control tower operators. I have paid little attention to the radar cases, since I am not a radar expert. The expert on Project Bluebook invariably ascribed all radar cases to malfunction or to anomalous propagation.* The Condon Report, however, contains the following remark about one such case: "This must remain as one of the most puzzling radar cases on record—and no conclusion is possible at this time. It seems inconceivable that an anomalous propagation echo would behave in the manner described, even if anomalous propagation had been likely at the time." [3]

Radar-visual cases offer more scope for study. The Lakenheath case (see Chapter 5, Appendix, Case 2), studied by the Condon committee, was left as unexplained with the remark, "In summary, this is the most puzzling and unusual case in the radar-visual files. The apparently rational, intelligent behavior of the UFO suggests a mechanical device of unknown origin as the most probable explanation of the sighting. However, in view of the inevitable fallibility of witnesses, more conventional explanations of this report cannot be entirely ruled out" (p. 164).[4] In actuality, a careful reading of the body of the Condon Report reveals as good a case for the scientific study of UFO's as could have been assembled by a group not initially conversant with the subject and with limited time and funds.

* The latter arises when meteorological conditions are such as to interfere with the normal straight-line propagation of radar waves, leading to erroneous interpretation of the radar results.

Some may be surprised that so considerable a body of UFO evidence exists. This is the crux of the problem: neither the active scientists nor the public have access to this information. Unfortunately, those who wish to learn about UFO's must get information from the "back fences" of literature—the pulp magazines, the sensational mystery or sex magazines. Until recently, there has not been in this country one scientific journal in which I could publish a well-documented UFO case, yet a recent bibliography of UFO literature ran to 400 pages.[5] The UFO has become a problem for the librarian sooner than it has for the scientist.

Consider the plight of serious UFO witnesses. I know that such exist because I have interviewed several hundred. Where can they go to report? Only the most naive would report to the Air Force even if Project Bluebook had not been discontinued. To report to the local police is scarcely better. Many witnesses have told me of the ridicule they met when they took that path. Besides, I have seen many police blotters. UFO reports are entered as "complaints."

The witness, if he wishes to report, must seek out the relatively few persons or organizations which will lend a sympathetic ear. My own mail brings me very good UFO reports, generally with a request for anonymity, but I have neither the time nor the funds to make proper investigations.

As I look back over my past twenty-one years' association with the UFO problem, I note that the intellectual climate today is enormously better for taking a good look at it than it was even a few years ago. This symposium is itself an example: it would have been impossible to have held it a year or two ago. And had I, earlier, attempted to call for a major investigation, I would have lost credibility and undoubtedly all possible future effectiveness.

In summary, then, my twenty-one years of monitoring of UFO reports has shown that reports of UFO observations remain after we delete the pronouncements of crackpots, visionaries, religious fanatics, and so forth. A large number of UFO reports are readily indentifiable by trained investigators as misperception of known objects and events. A small residue of UFO reports are *not* so identifiable. These come from such widely separated places as northern Canada, Australia, South America, and Antarctica. They are made by competent, responsible, psychologically normal people—in short, credible witnesses. These reports contain descriptive terms which *collectively* do *not* specify any *known physical* event, object, or process, and which do *not* specify any *known psychological* event or process. And, furthermore, they resist translation into terms that *do* apply to known physical and/or psychological events, objects, and processes. That is, as Goudge points out, translation would alter the meaning of the original report and hence effectively violate the methodological criteria governing the advance of science: namely (a) that it must be *possible* for new observational data to occur; that is, the existing conceptual framework of science, or the attitudes of scientists, must not rule out such new data *a priori;* and (b) the existing conceptual framework must allow new concepts, principles, and laws to be formulated to *interpret* and *explain* the new observational data.

Although I know of no hypothesis that adequately covers the mountainous evidence, this should not and must not deter us from following the advice of Schroedinger: to be curious, capable of being astonished, and eager to find out.

## NOTES

1. Thomas Goudge, personal letter to author.
2. E. U. Condon, *Scientific Study of Unidentified Flying Objects* (New York: Bantam Books, 1969).
3. *Ibid.*, p. 171.
4. *Ibid.*, p. 164.
5. Lynn Catoe, *UFO's and Related Subjects: An Annotated Bibliography*, Library of Congress, AFOSR 68–1656 (Washington, D.C., U.S. Government Printing Office, 1969).

# 5

## Science in Default: Twenty-two Years of Inadequate UFO Investigations

JAMES E. McDONALD

No scientifically adequate investigation of the UFO problem has been carried out during the entire twenty-two-year period between the first extensive wave of sightings of unidentified aerial objects in the summer of 1947 and the convening of this symposium. Despite continued public interest and frequent expression of public concern, only quite superficial examinations of the steadily growing body of unexplained UFO reports from credible witnesses have been conducted in this country and abroad. The latter point is highly relevant, since all evidence now points to the fact that UFO sightings exhibit similar characteristics throughout the world.

In charging inadequacy of all past UFO investigations, I speak not only from intimate knowledge of the past investigations, but also from three years of detailed personal research, involving interviews with more than five hundred witnesses in selected UFO cases, chiefly in the United States. In my opinion, the UFO problem, far from being the "nonsense problem" it has been labeled by many scientists, constitutes an area of extraordinary scientific interest.

The grave difficulty with essentially all past UFO studies has been that they either are devoid of substantial scientific content or they become lost amid the noise that tends to obscure the real signal in the UFO reports. The high percentage of reports of misidentified natural or technological phenomena (planets, meteors, and aircraft, above all) is not surprising, given all the circumstances surrounding the UFO problem. Yet such understandable and usually easily recognized instances of misidentification have all too often been seized upon as a sufficient explanation for *all* UFO reports, while the residue of far more significant reports (numbering now about one thousand) is ignored. I believe science is in default for having failed to mount any truly adequate studies of this problem. Unfortunately, the present climate of thinking since release of the study conducted under the direction of E. U. Condon at the University of Colorado will make it very difficult to secure more thorough investigations; yet my own examination of the problem forces me to call for just such new studies. I am enough of a realist to sense that, unless this AAAS symposium succeeds in making the scientific community aware of the seriousness of the UFO problem, little response to any call for new investigation is likely to appear.

In fact, the overall public and scientific response to the UFO phenomena is itself a matter of substantial sociopsychological interest. Prior to my own investigations, I would never have imagined the widespread reluctance to report an unusual and seemingly inexplicable event; yet that reluctance, and the reluctance of scientists to pay serious attention to the phenomena in question, are quite general. One regrettable result is the fact that the most credible of UFO witnesses are often those most reluctant to come forward with a report of the event they have witnessed. It is also regrettable that only a very small number of scientists have taken the time and trouble to search out the really puzzling reports which are almost lost among the trivial and nonsignificant UFO reports. The net result is that there still exists no general scientific recognition of the scope and nature of the UFO problem.

Within the federal government, official responsibility for UFO investigations has rested with the Air Force since early 1948, and unidentified aerial objects quite naturally fall within the area of Air Force concern. However, once it became clear in early 1949 that UFO reports did not involve advanced aircraft of a hostile foreign power, Air Force interest subsided to relatively low levels, with occasional temporary resurgence of interest following large waves of UFO reports, such as those of 1952, 1957, and 1965.

A most unfortunate pattern of press reporting developed by about 1953, in which the Air Force would assert that they had found "no evidence of anything defying explanation in terms of present-day science and technology" in their growing files of UFO reports. These statements to the public would have done little harm had they not been coupled systematically to press statements asserting that "the best scientific facilities available to the U.S. Air Force" had been and were being brought to bear on the UFO question. The assurances that substantial scientific competence was involved in Air Force UFO investigations had a seriously deleterious effect. Scientists who might otherwise have done enough checking to see that a substantial scientific puzzle lay in the UFO area were misled by these assurances into thinking that capable scientists had already made adequate study and found nothing. My own extensive checks have revealed so slight an amount of scientific competence in two decades of Air Force–supported investigations that I can only regard the repeated claim of solid scientific study of the UFO problem as a serious obstacle to progress toward elucidation.

Let me stress that this has not been part of some top-secret cover-up of investigations by Air Force or security agencies; I have found no basis for accepting that theory of why the Air Force has failed to respond to the many intriguing UFO reports coming from within its own ranks. In short, I see "grand foul-up" but not "grand cover-up."

Close examination of the level of investigation and scientific analysis involved in Project Sign (1948–1949), Project Grudge (1940–1952), and Project Bluebook (1953–1969), reveals that these were, viewed scientifically, almost meaningless investigations. Even during periods (e.g., 1952) of fairly active investigation of UFO cases, such slight scientific expertise was involved that there was never any real chance that the puzzling phenomena encountered in the most significant UFO cases would be elucidated. The panels, consultants, contractual studies, and so forth, that the Air Force has conducted over the past twenty-two years have brought almost negligible scientific scrutiny to bear on the UFO problem.

The Condon Report, released in January 1969 after about two years of Air Force–supported study, is, in my opinion, quite inadequate. The sheer bulk of the report, much of it "scientific padding," cannot conceal from anyone who studies it closely that it examines only a tiny fraction of the really puzzling UFO reports and that its scientific argumentation is often unsatisfactory. Of roughly ninety cases that it specifically confronts, more than thirty are conceded to be "unexplained." With so large a fraction of unexplained cases in a sample that is by no means limited to the truly puzzling cases (there is a large number of obviously trivial cases), it is far from clear how Dr. Condon felt justified in concluding "that further extensive study of UFO's probably cannot be justified in the expectation that science will be advanced thereby."

I shall cite a number of specific cases from the Condon Report which I regard as inadequately investigated and reported. One at Kirtland Air Force Base, November 4, 1957, involved observations of a wingless egg-shaped object that was seen hovering for about a minute over the airfield prior to its departure at a climb rate which was described to me as faster than that of any known jets, then or now. The principal witnesses were precisely the type of witnesses whose accounts warrant closest attention, since they were CAA tower observers who watched the UFO with binoculars. Yet, when I located these two men in the course of my own check of cases from the Condon Report, I found that neither of them had even been contacted by members of the Colorado Project! Both men were sure that they had been viewing a device with performance characteristics well beyond anything in present or foreseeable aeronautical technology. The two men gave me descriptions that were consistent and that fit the testimony given on November 6, 1957, when they were interrogated by an Air Force investigator. The Condon Report attempts to explain this case as a light aircraft that lost its way, came into the field area, and then left. This kind of explanation is often repeated in the Condon Report; yet it is wholly incapable of explaining the details of sightings. I will cite other instances where the investigations summarized in the Condon Report exhibit glaring deficiencies, and I suggest that there are enough significant unexplainable UFO reports within the Condon Report alone * to document the need for a greatly increased level of scientific study of UFO's.

* The following are UFO cases conceded to be unexplainable in the Condon Report and containing features of particularly strong scientific interest: Utica, N.Y., 6/23/55; Lakenheath, England, 8/13/56; Jackson, Ala., 11/14/56; Nor-

That a panel of the National Academy of Sciences could endorse this study is to me disturbing. I find no evidence that the Academy panel did any independent checking, and, to my knowledge, none of that eleven-man panel had any significant prior investigative experience in this area. I believe that this sort of endorsement hurts science in the long run and will ultimately prove an embarrassment to the Academy.

## APPENDIX : FOUR CASES

Four UFO cases may serve as specific illustrations of what I regard as serious shortcomings in the Condon Report and in the 1947–1969 Air Force UFO program. My principal conclusion is that scientific inadequacies in past years of investigations by Air Force Project Bluebook have *not* been remedied by publication of the Condon Report, and that there remain scientifically very important unsolved problems with respect to UFO's.

### Case 1. South-Central U.S., July 17, 1957

*Summary:* An Air Force RB-47, equipped with ECM (Electronic Countermeasures) gear and manned by six officers, was followed for a distance of well over 700 miles and for a time period of 1.5 hours, as it flew from Mississippi, through Louisiana and Texas, and into Oklahoma. The object was, at various times, seen visually by the cockpit crew as an intensely luminous light, followed by ground-radar, and detected on ECM monitoring gear aboard the RB-47. Of special interest

---

folk, Va., 8/30/57; RB-47 case, 9/19/57; Beverly, Mass., 4/22/66; Joplin, Mo., 1/13/67; Donnybrook, N.D., 8/19/66; Haynesville, La., 12/30/66; Colorado Springs, Colo., 5/13/67.

I take strong exception to the argumentation presented for the following UFO cases, considered explained in the Condon Report, and regard them as both unexplained and of strong scientific interest: Flagstaff, Ariz., 5/20/50; Washington, D.C., 7/19/52; Bellefontaine, O., 8/1/52; Haneda AFB, Japan, 8/5/52; Gulf of Mexico, 12/6/52; Odessa, Wash., 12/10/52; Continental Divide, N.M., 1/26/53; Seven Isles, Quebec, 6/29/54; Niagara Falls, N.Y., 7/25/57; Kirtland AFB, N.M., 11/4/57; Gulf of Mexico, 11/5/57; Peru, 12/30/66; Holloman AFB, 3/2/67; Kincheloe AFB, 9/11/67; Vandenberg AFB, 10/6/67; Milledgeville, Ga., 10/20/67.

in this case are several instances of simultaneous appearances and disappearances on all three physically distinct "channels," and a rapidity of maneuvers beyond the prior experience of the aircrew.

*Introduction*

This is an example of a category of UFO cases whose well-documented and puzzling details have never been adequately laid before a wide scientific audience. In my searches through Air Force archives, I have found many more, but this is one of the most interesting of the Air Force UFO cases that I examined.

Space limitations require that I present a rather condensed summary of this long and involved incident and its well-attested phenomena that defy easy explanation in terms of present-day science and technology.

The RB-47 was flying out of Forbes Air Force Base, Topeka, on a composite mission that included gunnery exercises over the Texas-Gulf area, navigation exercises over the open Gulf, and finally ECM exercises scheduled for the return trip across the south-central United States. Three of the six-man crew were electronic warfare officers manning ECM gear in the aft portion of the aircraft. One of the interesting features of this case is that electromagnetic signals of a distinctly radar-like nature were repeatedly monitored on two independent ECM channels during the extended period of contact with the unidentified object.

The first open scientific discussion of this case appears in the Condon Report (on pp. 136, 260, 877), as a result of its being mentioned, almost fortuitously, to investigators on the University of Colorado UFO Project. Whereas the discussion in the Condon Report is based on interviews with only three of the aircrew, the following is based on interviews with all six of the Air Force officers, and, perhaps more significantly, on examination of the original 1957 Air Force reports, which were never located by the University of Colorado group. The six men, with whom I have had a number of telephone discussions in pinning down the key points, are: Maj. Lewis D. Chase, pilot; 1st Lt. James H. McCoid, copilot; Thomas H. Hanley, navigator; John J. Provenzano, #1 monitor; Frank B. McClure, #2 monitor; Walter A. Tuchscherer, #3 monitor. I failed to ask for information on the 1957 ranks of the navigator and monitors. Chase, McClure, and Hanley are currently officers on active Air Force duty. Unfortunately, I have been unable to locate any personnel involved in the ground-radar observations that are an important part of the entire case.

The date assigned to this incident in the Condon Report is incorrect. It was based on the best recollections of Chase, McCoid, and McClure (the three who were interviewed by the Project), plus rough checks based on their personal flight logs. On that tentative basis, the incident was assigned a date of September 19-20, 1957. Actually the correction of the date to July 17, 1957, is one of the few really significant corrections made possible by locating the case file in the archives; most of the other recollections I obtained from interviews with the six members of the aircrew were closely supported by the detailed report prepared shortly after the incident in 1957. (I would stress that, contrary to suggestions that are often made, instances such as this, where I have had the opportunity to secure independent contemporaneous accounts of UFO sightings of extremely unusual nature that left vivid impressions in the minds of the witnesses, tend to reveal that time does not seriously distort either major features or many finer details of the original incidents. Rather, it would be my evaluation that the more unusual the basic experience the more reliable the recollective account tends to be.)

Actually, the case file for this incident (like that of many other scientifically significant UFO cases) is less than complete. It is possible that some of the original investigative records were sent, not to Project Bluebook, but to Air Defense Command intelligence units, as several of the RB-47 crew indicated to me. Checks by members of the Colorado Project team revealed that all records in ADC intelligence archives are routinely destroyed after a period that has usually been three years. The men described quite detailed interrogation, which does not appear to be reflected in the Bluebook case file; a number of significant points on which there was general agreement by all six of the officers are scarcely mentioned. On the other hand, there is relevant information in the file as to precise times, locations, and so on, and the file does have the virtue of representing a summary-account prepared while all of the details were fresh in the minds of the crew.*

* In addition to my direct interviews with the crew, the account that follows is based upon a three-page TWX filed from the 745th ACWRON, Duncanville,

*Initial ECM Contact*

The account in the Condon Report conveys no impression of the extended geographical range and duration in time of this incident; yet both Chase, who first gave a detailed account of the events to investigators on the Colorado Project, and McClure, whom they interviewed, were quite definite in pointing out to me that the initial ECM contact was made in southern Mississippi and that the final contact was lost in south-central Oklahoma. Those recollective statements are supported by the contemporaneous account prepared by the Wing Intelligence Officer. From my interviews with the six officers, I obtained estimates of time duration that all fell in the range of about an hour or more; in the contemporaneous records the time lapse between the first *visual* contact (1010Z *) and the final ECM contact (1140Z) is 1.5 hours. And, as will now be made clear, an important portion of the incident took place even prior to first visual contact at 1010Z. The original TWX from the Duncanville ground-radar unit gives the distance over which the RB-47 was followed by the unidentified object as 520 miles. However, from the more detailed subsequent intelligence report, it is clear that the total distance was close to 800 miles.

Before describing the first ECM contact, it is necessary to explain briefly the nature of the ECM gear involved in this case. (Details are no longer classified, although all of the basic case file documents were initially "secret." They were declassified January 29, 1969, by authorization of Lt. Col. H. Quintanilla, the officer in charge of Project Bluebook; but the fact that the declassification date is twenty days after the final re-

---

Texas, at 1557Z on 7/17/57 and a four-page case summary prepared by E. T. Piwetz, Wing Intelligence Officer, 55th Reconnaissance Wing, Forbes AFB, and transmitted to ADC Hq, Ent AFB, Colorado, in compliance with a request of 8/15/57 from Col. F. T. Jeep, Director of Intelligence, ADC. That summary, plus a twelve-page Airborne Observer's Data Sheet, was forwarded on 10/17/57 from ADC to Bluebook, evidently the first notification they received concerning this case. The Data Sheet (AISOP 2) was prepared by Major Chase on 9/10/57 and contains a number of points of relevance not covered in other parts of the case file.

* "1010Z" stands for 10:10 A.M. Greenwich Z-zone time, generally used in military reports. Central Standard Time was then 4:10 A.M.

lease of the Condon Report should not be taken to imply that this case file was withheld from the Colorado Project. Other classified reports were made available, despite their classification; all evidence points to the conclusion that ignorance of the exact date and failure to search far enough in each direction from the estimated date led to the failure to locate this file for discussion in the Condon Report.) This RB-47 had three passive direction-finding (DF) radar monitors for use in securing coordinate information and pulse characteristics on enemy ground-based radar. The #2 monitor, manned by McClure, was an ALA-6 DF receiver with back-to-back antennas in a housing on the belly of the RB-47 near the tail, spun at 150 or 300 rpm as it scanned an azimuth. (Note that this implies ability to scan at 5 / sec past a fixed ground radar in the distance.) Inside the aircraft, the signals from the ALA-6 were processed in an APR-9 radar receiver and an ALA-5 pulse analyzer. All subsequent references to the #2 monitor imply that system. The #1 monitor, manned by Provenzano, was an APD-4 DF system, with a pair of antennas permanently mounted on either wingtip. The #3 monitor, manned by Tuchscherer, was not involved and will not be described.

These DF receivers are *not* radars and do not emit a signal for reflection off a distant target. They only listen to incoming radar signals and perform signature analyses. When receiving a distant radar set's signal, the scope displays a pip or strobe at an azímuthal position corresponding to the relative bearing in the aircraft coordinate system. For the case of a fixed ground radar approached from one side, the strobe is initially seen in the upper part of the scope and moves *downscope,* a point to be carefully noted in interpreting the following discussion.

Having completed the navigational exercises over the Gulf, Chase headed north across the Mississippi coastline, flying at an altitude of 34,500 feet, at about Mach 0.75 (258 kt IAS = 500 mph TAS). Shortly after they crossed the coast near Gulfport, McClure detected on the #2 monitor a signal painting at their 5 o'clock position (aft of the starboard beam). It looked to him like a legitimate ground-radar signal, and, upon noting that the strobe was moving *upscope,* McClure tentatively decided that it must be a ground radar off to their northwest, painting with 180° ambiguity for some electronic reason. But when the strobe, after sweeping upscope on the starboard side, crossed the flight path of the RB-47 and proceeded to move *downscope* on the port side, McClure said he gave up the hypothesis of 180° ambiguity as incapable of explaining such behavior. Fortunately, he had examined the signal characteristics on his ALA-5 pulse analyzer before the signal left his scope on the port side aft. In discussing it with me, his recollection was that the frequency was near 2800 mcs, and he recalled that what was particularly odd was that it had a pulse width and pulse repetition frequency (PRF) much like that of a typical S-band ground-based search radar. He even recalled that there was a simulated scan rate that was normal. Perhaps because of the strong similarities to ground-based sets such as the CPS-6B, widely used at that time, McClure did not, at that juncture, call this signal to the attention of anyone else in the aircraft. The #1 monitor was not working the frequency in question, it later developed. The #3 monitor was incapable of working the frequency in question, McClure and the others indicated to me.

I next quote information transcribed from the summary report prepared by the Wing Intelligence Officer, COMSTRATRECONWG 55, Forbes Air Force Base, concerning this part of the incident that involved this aircraft (call sign "Lacy 17"):

ECM reconnaissance operator #2 of Lacy 17, RB-47H aircraft, intercepted at approximately Meridian, Mississippi, a signal with the following characteristics: frequency 2995 MC to 3000 MC; pulse width of 2.0 microseconds; pulse repetition frequency of 600 cps; sweep rate of 4 rpm; vertical polarity. Signal moved rapidly up the D/F scope indicating a rapidly moving signal source; i.e., an airborne source. Signal was abandoned after observation.

*Initial Visual Contact*

If nothing further had occurred on that flight to suggest that some unusual object was in the vicinity of the RB-47, McClure's observations undoubtedly would have been quickly forgotten even by him. He was puzzled, but at that point still inclined to think that it was some electronic difficulty. The flight plan called for a turn to the west in the vicinity of Meridian and Jackson, Mississippi, with subsequent planned exercises wherein the aircraft did simulate ECM runs against known ground-radar units. The contemporary records confirm what Chase and

McCoid described to me far more vividly and in more detail concerning the unusual events that ensued. They turned into a true heading of 265°, still at Mach 0.75 at 34,500 feet. At 1010Z (0410 CST), Major Chase, in the forward seat, spotted what he first thought were the landing lights of another jet coming in fast from near his 11 o'clock position at, or perhaps a bit above, the RB-47's altitude. He called McCoid's attention to it, noted absence of any navigational lights, and, as the single intense bluish white light continued to close rapidly, he used the intercom to alert the rest of the crew to be ready for sudden evasive maneuvers. But before he could attempt evasion, he and McCoid saw the brilliant light almost instantaneously change direction and flash across their flight path from port to starboard at an angular velocity that Chase told me he had never seen matched in all of his twenty years of flying, before or after that incident. The luminous source had moved with great rapidity from their 11 o'clock to about their 2 o'clock position and then blinked out.

The Airborne Observer's Data Sheet filled out by Chase as part of the postinterrogation gives the RB-47 position at the time of that 1010Z first visual contact as 32°00′N, 91°28′W, which puts it near Winnsboro in east-central Louisiana (Point C on the map, figure 5-1).

Again, it is to be noted that the descriptions I obtained in my 1969 interviews with these officers are closely supported by the original intelligence report:

> At 1010Z aircraft comdr first observed a very intense white light with light blue tint at 11 o'clock from his aircraft, crossing in front to about 2:30 o'clock position, copilot also observed passage of light to 2:30 o'clock where it apparently disappeared.

*Actions over Louisiana-Texas Area*

Immediately after the luminous source blinked out, Chase and McCoid began talking about it on the interphone, with the alerted crew listening in. McClure now mentioned the unusual signal he had received on his ALA-6 back near Gulfport, set his #2 monitor to scan at about 3000 mcs to see what might show up. He found he was getting a strong 3000 mcs signal from about their 2 o'clock position, the rela-

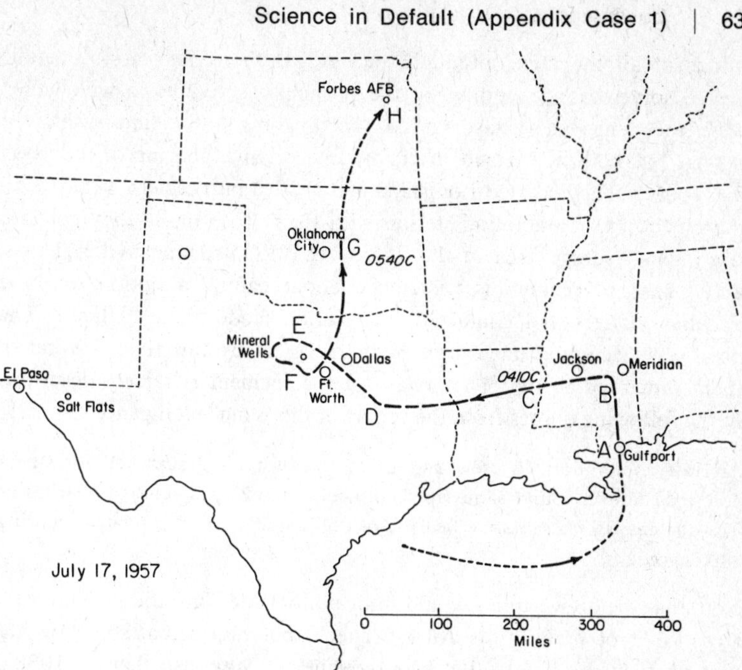


*Figure 5-1.* Map of July 17, 1957, UFO episode.

tive bearing at which the unknown luminous source had blinked out moments earlier.

Provenzano told me that immediately afterward they checked out the #2 monitor on other known ground-radar stations, to be sure that it was not malfunctioning; it appeared to be in perfect working order. He then tuned his own #1 monitor to 3000 mcs and also got a signal from the same bearing. There remained, of course, the possibility that, by chance, this signal was from a real radar on the ground in that relative direction. But as the minutes then went by and the RB-47 continued westward at about 500 mph, the relative bearing of the 3000 mcs source out in the dark did *not* move downscope on the monitors, as should have occurred with any ground radar, but instead kept up with the RB-47, holding a fixed relative bearing.

I found these and ensuing portions of the episode still vivid in the minds of all the men, although their recollections for various details varied somewhat, depending on the particular activities in which they were then engaged. Chase varied speed, going to maximum allowed power, but nothing seemed to change the relative bearing of the 3000 mcs source. They crossed Louisiana and headed into eastern Texas, with the object still maintaining station with them. Eventually they got into the radar-coverage area of the 745th ACWRON, Duncanville, Texas, and Chase overcame his reluctance about calling attention to these peculiar matters and contacted that station (code name "Utah"). The crew was becoming uneasy about the incident by this time, several of them remarked to me. That phase of the incident is tersely described in the following quotes from the report of the Wing Intelligence Officer:

Aircraft comdr notified crew and ECM operator Nr 2 searched for signal described above, found same approximately 1030Z at a relative bearing of 070 degrees; 1035Z, relative bearing of 068 degrees; 1038Z, relative bearing 040 degrees.

Note that the above times would indicate that McClure did not immediately think of making his ALA-6 check, but rather that some twenty minutes went by before that was thought of. Note also that by 1038Z the unknown source of the 3000 mcs radarlike signal was moving *up-scope* relative to the 500 mph RB-47.

At 1039Z aircraft comdr sighted huge light which he estimated to be 5000 below aircraft at about 2 o'clock. Aircraft altitude was 34,500 ft, weather perfectly clear. Although aircraft comdr could not determine shape or size of object, he had a definite impression light emanated from top of object.

At 1040Z ECM operator #2 reported he then had two signals at relative bearings of 040 and 070 degrees. Aircraft comdr and co-pilot saw these two objects at the same time with same red color. Aircraft comdr received permission to ignore flight plan and pursue object. He notified ADC site Utah and requested all assistance possible. At 1042Z ECM #2 had one object at 020 degrees, relative bearing.

In my interviews with the aircrew, I found differences between their recollections of some of these points. McCoid recalled that the luminous source occasionally moved abruptly from starboard to port side and back again. Chase recalled that they had contacted Utah (his recollection was that it was Carswell GCI, however) prior to some of the above events and that Utah was ground-painting the target during the time it moved upscope and reappeared visually. As will be seen below, the contemporary account makes fairly clear that Utah was not painting the unknown until a bit later, after it had turned northwestward and passed between Dallas and Fort Worth. Chase explained to me that he got FAA clearance to follow it in that off-course turn, and indicated that FAA got all jets out of the way to permit him to maintain pursuit. The intelligence summary continues:

At 1042Z ECM #2 had one object at 020 degrees relative bearing. Aircraft comdr increased speed to Mach 0.83, turned to pursue, and object pulled ahead. At 1042.5Z ECM #2 again had two signals at relative bearings of 040 and 070 degrees. At 1044Z he had a single signal at 050 degrees relative bearing. At 1048Z ECM # 3 was recording interphone and command position conversations.

ADC site requested aircraft to go to IFF Mode III for positive identification and then requested position of object. Crew reported position of object as 10NM northwest of Ft. Worth, Texas, and ADC site Utah immediately confirmed presence of objects on their scopes.

At approximately 1050Z object appeared to stop, and aircraft overshot. Utah reported they lost object from scopes at this time, and ECM #2 also lost signal.

Chase, in reply to my questions, said he recalled that there was simultaneity between the moment when he began to sense that he was getting closure at approximately the RB-47 speed and the moment when Utah indicated that their target had stopped on their scopes. He said he veered a bit to avoid colliding with the object, not then being sure what its altitude was relative to the RB-47, and then found that he was coming over the top of it as he proceeded to close. At the instant that it blinked out visually and disappeared simultaneously from the #2 monitor and from the radar scopes at site Utah, it was at a depression angle relative to his position of something like 45 degrees.

Chase put the RB-47 into a port turn in the vicinity of Mineral Wells, Texas (Point E on accompanying map), and he and McCoid looked over their shoulders to try to spot the luminous source again. All of the men recalled the near-simultaneity with which the object blinked on again visually, appeared on the #2 scope, and was again skin-painted by ground radar at site Utah. The 1957 report describes these events as follows:

> Aircraft began turning, ECM #2 picked up signal at 160 degrees relative bearing, Utah regained scope contact, and aircraft comdr regained visual contact. At 1052Z ECM #2 had signal at 200 degrees relative bearing, moving up his D/F scope. Aircraft began closing on object until the estimated range was 5 NM. At this time object appeared to drop to approximately 15,000 feet altitude, and aircraft comdr lost visual contact. Utah also lost object from scopes.
>
> At 1055Z in the area of Mineral Wells, Texas, crew notified Utah they must depart for home station because of fuel supply. Crew queried Utah whether a CIRVIS report had been submitted, and Utah replied the report had been transmitted. At 1057Z ECM #2 had signal at 300 degrees relative bearing, but Utah had no scope contact. At 1058Z aircraft comdr regained visual contact of object approximately 20 NM northwest of Ft. Worth, Texas, estimated altitude 20,000 ft at 2 o'clock from aircraft.

Chase added further details on this portion of the events, stating that he requested and secured permission from Utah to dive on the object when it was at lower altitude. He did not recall the sudden descent that is specified in the contemporary account, and there are a number of other minor points in the intelligence report that were not recollected by any of the crew. He told me that, when he dove from 35,000 feet to approximately 20,000 feet, the object blinked out, disappeared from the Utah ground-radar scopes, and disappeared from the #2 monitor, all at the same time. McClure recalled that simultaneous disappearance, too. It should be mentioned that the occasional appearance of a second visual and radar-emitting source was not recalled by any of the officers when I interviewed them in 1969. One gathers from the intelligence report that the appearance of two objects was limited to a short time period, while the aircraft was over east-central Texas.

*Actions over Texas-Oklahoma Area*

McCoid recalled that, at about this stage of the activities, he was becoming a bit worried about excess fuel consumption resulting from use of maximum allowed power, plus a marked departure from the initial flight plan. He advised Chase that fuel limitations would necessitate a return to the home base at Forbes AFB, so they soon headed north from the Fort Worth area. McClure and Chase recalled that the ALA-6 system again picked up a 3000 mcs signal on their tail, once they were northbound from Fort Worth, but there was some variance in their recollections as to whether the ground radar concurrently painted the object. McCoid was unable to fill in any of those details. Fortunately, the 1957 intelligence report summarizes further events in this part of the flight, as they moved northward into Oklahoma:

> At 1120Z aircraft took up heading for home station. This placed area of object off the tail of aircraft. ECM #2 continued to [get] D/F signal of object between 180 and 190 degrees relative bearing until 1140Z, when aircraft was approximately abeam Oklahoma City, Oklahoma. At this time, signal faded rather abruptly. 55 SRW DOI [55th Strategic Reconnaissance Wing, Director of Intelligence] has no doubt the electronic D/F's coincided exactly with visual observations by aircraft comdr numerous times, thus indicating positively the object being the signal source.

It was Chase's recollection that the object was with them only into southern Oklahoma; Hanley recalled that it was with them all the way to the Oklahoma City area; the others remembered only that it was there for some indefinite distance on the northbound leg between Fort Worth and Topeka, their home base.

*Bluebook and Condon Report Evaluations*

The records indicate that Project Bluebook received summary information on this incident from ADC on October 25, 1957 (over three months after occurrence of the event). An initial evaluation by V. D. Bryant of the Electronics Branch, Wright-Patterson, AFB, dated October 30, 1957, comments as follows:

> This report is difficult to evaluate because there is such a mass of evidence which tends to all tie in together to indicate the presence of a physical object or UFO. With the exception of rather abrupt disappearance of returns on the electronic equipment, an indication that the object travelled at relatively high speed, there are no abnormal electronic indications such as are usually present in reports of the type—extreme speeds, abrupt changes

of course, etc. These abnormal indications are usually the basis for considering anomalous propagation, equipment malfunction, etc., as responsible for the "sightings."

The electronic data is unusual in this report in that radar signals (presumably emanating from the "object") were picked up. These intercepted signals have all the characteristics of ground-radar equipment, and in fact are similar to the CPS-6B. This office knows of no S-band airborne equipment having the characteristics outlined. Since the type equipment on the ground (at "Utah") is not known, and since there are no "firm" correlations between the ground intercept and the sightings from the aircraft, it is impossible to make any determination from the information submitted. On the other hand, it is difficult to conclude that nothing was present, in the face of the visual and other data presented.

The only other material in the case file that constitutes anything like a Bluebook analysis is an undated "Brief Summary" of the case by Capt. G. T. Gregory, then Bluebook Officer. The bulk of his summary is a recapitulation of information already presented; the important statement is the final paragraph: "In joint review with the CAA of the data from the incident, it was definitely established by the CAA that object observed in the vicinity of Dallas and Ft. Worth was an airliner."

The reader who examines for himself the analysis presented in the Condon Report will find some confusion between the three different sections wherein this case is discussed. At one point there is speculation that the visual observations might have been due to oil-well flares in the vicinity of Oklahoma City; in another, it is speculated that perhaps a thin inversion layer near the flight altitude might have caused some kind of an optical distortion of jet airliner headlights; in still another part of the report, the ground-radar observations seem to be attributed to anomalous propagation. However, in the end, the RB-47 case is assigned to the "unidentified" category in the Condon Report.

*Discussion*

The simultaneity of the appearance and disappearance of the visual signal, the ECM-monitored electromagnetic signal, and the skin-paint return seen on the ground radar, taken together with the extremely unconventional maneuvers exhibited by this source, which was able to outfly an RB-47 (and even able to fly an "orbit" around it at one time), would seem sufficient to exclude all of the hypotheses considered by Bluebook and in the Condon Report.

The case is now carried in the official Bluebook files as "Identified as American Airlines Flight 655." That evaluation was Captain Gregory's 1957 conclusion. In the case file is a sheet containing the following brief comment on the American Airlines incident: "July 17—50 miles E of El Paso, Texas—3:30 A.M. (MST) Amer. Airlines Flight #655 almost collides with huge green UFO! (Shot E.) (Fireballs mounting)."

Another item in the same file adds further unofficial comment:

> The American Airlines DC-6 air coach with 85 aboard narrowly averted collision near Salt Flats, Texas, in the pre-dawn darkness of July 17, 1957. Capt. Ed Bachner dived the airliner from its 14,000 ft altitude when he saw a green light ahead. Ten passengers were injured when thrown from their seats. Though the weather was clear, the crew said the other aircraft appeared without warning.

Official comment on this incident is contained in a letter dated November 8, 1957, from Roy Keeley, Director, Office of Flight Operations and Air Worthiness, Civil Aeronautics Administration, addressed to Brig. Gen. Harold E. Watson, Air Technical Intelligence Center, Wright-Patterson AFB, Ohio. The relevant paragraph is the following:

> The second incident mentioned occurred on July 17, 1957, near El Paso, Texas, and involved American Airlines Flight #655. Investigation of this incident definitely established the fact that the unidentified flying object was American Airlines Flight #966, which had previously departed from El Paso, Texas, en route to Dallas, Texas.

Salt Flats, Texas, 50 miles east of El Paso, is approximately 450 miles west of the Fort Worth area. To equate that evidently explained incident with the complex sequence of events described above and involving the RB-47 during 1.5 hours of flight from southern Mississippi into Oklahoma, is less than reasonable, yet that is the official Bluebook explanation.

Similarly, none of the conjectures advanced in the Condon Report to give tentative explanation to this incident are at all adequate. The reader of that report is, regrettably, given only an inadequate account of the real nature of the events, their duration, and geographical extent. Without further comment and examination of a number of other quite unsatisfactory hypotheses, I am forced to conclude that this interesting sequence of events on July 17, 1957, is best described as involving an unidentified flying object having some quite interesting performance characteristics and electromagnetic capabilities. I hope to describe more adequately the basis for such a conclusion elsewhere.

## Case 2. Lakenheath and Bentwaters RAF/USAF, England, August 13–14, 1956

*Summary:* Unidentified objects are observed by USAF and RAF personnel over a period of five hours, and involving ground-radar, airborne-radar, ground-visual, and airborne-visual sightings of high-speed unconventionally maneuvering objects in the vicinity of two RAF stations at night. It is Case 2 in the Condon Report and is there conceded to be unexplained.

### *Introduction*

This case illustrates the fact that many scientifically intriguing UFO reports have lain in USAF Bluebook files for years without knowledge thereof by the scientific community. It represents a large subset of UFO cases in which all of the observations stemmed from military sources and which, had there been serious and competent scientific interest operating in Project Bluebook, could have been very thoroughly investigated while the information was fresh. It illustrates that the actual levels of investigation were entirely inadequate in even as unexplainable and involved cases as this one. It illustrates the uncomfortably incomplete and internally inconsistent features that one encounters in almost every report of its kind in the USAF Bluebook files at Wright-Patterson AFB, features attesting to the dearth of scientific competence in the Air Force UFO investigations over the past twenty-two years. It illustrates, when the original files are carefully studied and compared with the discussion thereof in the Condon Report, shortcomings in the report's presentation and critique of this and other cases. Finally, I believe this case is an example of those cases conceded to be unexplainable by the Condon Report that argue need for much more extensive and more thorough scientific investigation of the UFO problem, a need negated in the Condon Report and in the Academy endorsement thereof.

My discussion of this case is based upon the thirty-page Bluebook case file, plus certain other information presented in the Condon Report. This "Lakenheath case" was not known outside USAF circles before publication of the Condon Report. The names of military personnel involved are not given in the report. (Witness names, dates, and locales are deleted from all of the main group of fifty-nine cases, seriously impeding independent scientific check of case materials.) I secured copies of the case file from Bluebook, but all names of military personnel involved in the incident were cut out of the Xerox copies prior to releasing the material to me. Hence I have been unable to interview the key witnesses personally. However, there is no indication that anyone on the Colorado Project did any personal interviews, either; so it would appear I have had access to the same basic data used in the Condon Report's treatment of this extremely interesting case.

In the Lakenheath case, deletion of locality names creates confusion for the reader: three distinct RAF stations figure in the incident, and the discharged noncommissioned officer from whom the Colorado Project first received word of this UFO episode confused the names of two of those stations in his own account. That, plus other reportorial deficiencies in the presentation of the Lakenheath case, will almost certainly have concealed its real significance from most readers of the Condon Report.

Unfortunately, the basic Bluebook file is itself about as confusing as most Bluebook files on UFO cases. I shall attempt to mitigate as many of those difficulties as I can in the following, by putting the account into better overall order than in the Condon Report treatment.

*General Circumstances*

The entire episode lasted from about 2130Z,* August 13, to 0330Z, August 14, 1956; thus this is a nighttime case. The events occurred in east-central England, chiefly in Suffolk. The initial reports centered around Bentwaters RAF Station, located about six miles east of Ips-

* "2130Z" stands for 9:30 P.M. Greenwich zone-Z time, or local standard time in England, used throughout this case.

wich, near the coast, whereas much of the subsequent action centers around Lakenheath RAF Station, located some twenty miles northeast of Cambridge. Sculthorpe RAF Station also figures in the account, but only to a minor extent; it is near Fakenham, in the vicinity of The Wash. GCA (Ground Controlled Approach) radars at two of these three stations were involved in the ground-radar sightings, as was an RATCC (Radar Traffic Control Center) radar unit at Lakenheath. The USAF noncom who wrote to the Colorado Project about this incident was a Watch Supervisor on duty at the Lakenheath RATCC unit that night. His detailed account is reproduced in the Condon Report (pp. 248–251). The report comments on "the remarkable accuracy of the account of the witness as given in [his reproduced letter], which was apparently written from memory 12 years after the incident." I would concur, but would note that, had the Colorado Project only investigated more such striking cases of past years, it would have found many other witnesses in UFO cases whose vivid recollections often match verifiable contemporary accounts surprisingly well. My experience has been that, in multiple-witness cases where one can evaluate the consistency of recollections, the more unusual and inexplicable the original UFO episode, the more it impressed upon the several witnesses' memories a meaningful and still-useful pattern of relevant recollections. Doubtless, another important factor operates: the UFO incidents that are most striking and most puzzling probably have been discussed by the key witnesses enough times that their recollections have been thereby reinforced in a useful way.

The only map for this case in the Condon Report is based on a sketch made by the noncom who alerted them to the case. It is misleading, for Sculthorpe is shown fifty miles east of Lakenheath, whereas it actually lies thirty miles north-northeast. The map does not show Bentwaters at all; it is some forty miles east-southeast of Lakenheath. Even as basic facts as those locations do not appear to have been ascertained by those who prepared the discussion of this case in the Condon Report, which is most unfortunate, yet not atypical.

That this incident was subsequently discussed by many Lakenheath personnel was indicated to me by a chance event. In the course of my investigations of another radar UFO case from the Condon Report, that of September 11, 1967, at Kincheloe AFB, I found that the radar operator there had previously been stationed with the USAF detachment at Lakenheath and learned of the events secondhand because they were still being discussed by radar personnel when he arrived many months later.

*Initial Events at Bentwaters, 2130Z to 2200Z*

One of the many unsatisfactory aspects of the Condon Report is its frequent failure to put before the reader a complete account of the UFO cases it purports to analyze scientifically. In the present instance, the report omits all details of *three* quite significant radar sightings made by Bentwaters GCA personnel prior to their alerting the Lakenheath GCA and RATCC groups at 2255 LST. This omission is certainly not due to correspondingly slight mention in the original Bluebook case file; rather, the Bentwaters sightings actually receive *more* Bluebook attention than the subsequent Lakenheath events. Hence, I do not see how such omissions in the Condon Report can be justified.

(1) *First radar sighting, 2130Z.* Bentwaters GCA operator, A/2c—— (I shall use a blank to indicate the names razor-bladed out of my copies of the case file) reported picking up a target 25–30 miles ESE, which moved at very high speed on constant 295° heading across his scope until he lost it 15–20 miles northwest of Bentwaters. In the Bluebook file, A/2c——is reported as describing it as a strong radar echo, comparable to that of a typical aircraft, until it weakened near the end of its path across his scope. He is quoted as estimating a speed of the order of 4,000 mph, but two other cited quantities suggest even higher speeds. A transit time of 30 seconds is given, and if one combines that with the reported range of distance traversed, 40–50 miles, a speed of about 5,000–6,000 mph results. Finally, A/2c____stated that it covered about 5–6 miles per sweep of the AN/MPN-11A GCA radar he was using. The sweep period for that set is given as 2 seconds (30 rpm), so this yields an even higher speed estimate of about 9,000 mph. (This figure derives from the type of observation most likely to be accurate, in my opinion. The displacement of a series of successive radar blips on a surveillance radar such as the MPM-11A can be estimated to perhaps a mile or so with little difficulty when the operator has as large a number

of successive blips to work with as is here involved. Nevertheless, it is necessary to regard the speed as quite uncertain, though presumably in the range of several thousand miles per hour and hence not associable with any conventional aircraft, nor with still higher-speed meteors.)

(2) *Second radar sighting, 2130–2155Z.* A few minutes after the preceding event, T/Sgt____picked up on the same MPN-11A a group of 12–15 objects about 8 miles southwest of Bentwaters. In the report to Bluebook, he pointed out that "these objects appeared as normal targets on the GCA scope and . . . normal checks made to determine possible malfunctions of the GCA radar failed to indicate anything was technically wrong." The dozen or so objects were moving northeast together at speeds ranging between 80 and 125 mph, and "the 12 to 15 unidentified objects were preceded by 3 objects which were in a triangular formation with an estimated 1000 feet separating each object in this formation." The dozen objects to the rear "were scattered behind the lead formation of 3 at irregular intervals with the whole group simultaneously covering a 6 to 7 mile area," the official report notes.

Consistent radar returns came from this group during their 25-minute movement from the point at which they were first picked up, 8 miles southwest, to a point about 40 miles northeast, of Bentwaters, their echoes decreasing in intensity as they moved off to the northeast. When the group reached a point some 40 miles northeast, they all appeared to converge to form a single radar echo whose intensity is described as several times larger than a B-36 return under comparable conditions. Then motion ceased, while this single strong echo remained *stationary* for 10–15 minutes. Then it resumed motion to the northeast for 5–6 miles, stopped again for 3–5 minutes, and finally moved northward and off the scope.

(3) *Third radar sighting, 2200Z.* Five minutes after the foregoing formation moved offscope, T/Sgt____detected an unidentified target about 30 miles east of the Bentwaters GCA station, and tracked it in rapid westward motion to a point about 25 miles west of the station, where the object "suddenly disappeared off the radar screen by rapidly moving out of the GCA radiation pattern," according to his interpretation of the event. Here, again, we get discordant speed information, for T/Sgt____gave the speed only as being "in excess of 4,000 mph,"

whereas the time-duration of the tracking, given as 16 seconds, implies a speed of 12,000 mph, for the roughly 55-mile track-length reported. Nothing in the Bluebook files indicates that this discrepancy was investigated further or even noticed, so one can say only that the apparent speed was far above that of conventional aircraft.

(4) *Other observations at Bentwaters.* A control tower sergeant, aware of the concurrent radar tracking, noted a light "the size of a pinhead at arm's length," at about 10° elevation to the SSE. It remained there for about one hour, intermittently appearing and disappearing. Since Mars was in that part of the sky at that time, a reasonable interpretation is that the observer was looking at that planet.

A T-33 of the 512th Fighter Interceptor Squadron, returning to Bentwaters from a routine flight at about 2130Z, was vectored to the northeast to search for the group of objects being tracked in that sector. Their search, unaided by airborne radar, led to no airborne sighting of any aircraft or other objects in that area, and after about 45 minutes they terminated search, having seen only a bright star in the east and a coastal beacon as anything worth noting. The Bluebook case file contains 1956 USAF discussions of the case that make a big point of the inconclusiveness of the tower operator's sighting and the negative results of the T-33 search, but say no more about the much more puzzling radar-tracking incidents than to stress that they were of "divergent" directions, intimating that this somehow put them in the category of anomalous propagation (AP), which scarcely follows. None of the three cited radar sightings exhibits any features typical of AP echoes. The winds over the Bentwaters area are given in the file. They jump from the surface level (winds from 230° at 5–10 knots) to the 6,000-foot level (260°, 30 knots), and then hold at a steady 260° up to 50,000 feet, with speeds rising to a maximum of 90 knots near 30,000 feet. Even if one sought to invoke the highly dubious Borden-Vickers hypothesis (moving waves on an inversion surface*), not even the slowest of the tracked echoes (80–125 mph) could be accounted for, nor is it even clear that the direction would be explainable. Furthermore, the strength of the individual echoes (stated as comparable to normal aircraft re-

* R. C. Borden and C. K. Vickers of the Civil Aeronautics Authority made this suggestion in a 1953 study described on pages 156–157 of the Condon Report.

turns), the merging of the 15 or so into a single echo, the two intervals of stationarity, and final motion offscope at a direction about 45° from the initial motion, are all wholly unexplainable in terms of AP in these 2130–2155Z incidents. The extremely high-speed westward motion of single targets is even further from any known radar anomaly associated with disturbed propagation conditions. Blips that move across scopes from one sector to the opposite, in steady heading at steady apparent speed, correspond neither to AP nor to internal electronic disturbances. Nor could interference phenomena fit such observed echo behavior. Thus, this 30-minute period, 2130-2200Z, embraced three distinct events for which no satisfactory explanation exists. That these three events are omitted from the discussions in the Condon Report is unfortunate, for they serve to underscore the scientific significance of subsequent events at both Bentwaters and Lakenheath stations.

*Comments on Reporting of Events after 2255Z, August 13, 1956*

The events summarized above were communicated to Bluebook by Capt. Edward L. Holt of the 81st Fighter-Bomber Wing stationed at Bentwaters, as Report No. IR-1-56, dated August 31, 1956. All events occurring after 2200Z, on the other hand, were communicated to Project Bluebook via an earlier, lengthy teletype transmission from the Lakenheath USAF unit, sent out in the standard format of the report-form specified by regulation AFR200-2. Two teletype transmissions, dated August 17 and 21, 1956, identical in basic content, were sent from Lakenheath to Bluebook. The Condon Report presents the content of that teletype report in full, (pp. 252–254), except for deletion of all names and localities and omission of one important item to be noted later here. However, most readers will be entirely lost because what is presented is actually a set of answers to questions that are not stated! The version of AFR200-2 appearing in the report's Appendix B (pp. 819–826, there identified by its current designation, AFR80-17) would provide the reader with the standardized questions needed to translate much of the otherwise confusing array of answers on pp. 252–254. For that reason, plus others, many readers will almost certainly be greatly (and unnecessarily) confused on reading this important part of the Lakenheath account in the Condon Report.

The confusion, unfortunately, does not wholly disappear upon laboriously matching questions with answers, for it has long been one of the salient deficiencies of the USAF program of UFO report collection that the format of AFR200-2 (and its sequel AFR80-17) is usually only barely adequate and (especially for complex episodes such as that involved here) often too narrow in scope to set out clearly and chronologically all the events that may be of scientific significance. Anyone who has studied many Bluebook reports in the AFR200-2 format, dating back to 1953, will be aware of this difficulty. Failure to carry out even modest followup investigations and incorporate findings thereof into Bluebook case files leaves most intriguing Bluebook UFO cases full of unsatisfactorily answered questions. But those deficiencies do not, in my opinion, prevent the careful reader from discerning that large numbers of those UFO cases carry significant scientific implications, implications of an intriguing problem that has been largely unexamined in past years.

*Initial Alerting of Lakenheath GCA and RTCC*

The official files give no indication of any further UFO radar sightings by Bentwaters GCA between 2200 and 2255Z, when another fast-moving target was picked up 30 miles east of Bentwaters, heading almost due west at a speed given as "2,000–4,000 mph." It passed almost directly over Bentwaters, disappearing from their GCA scope for the usual beam-angle reasons when it came within 2–3 miles (the Condon Report intimates that this close-in disappearance is diagnostic of anomalous propagation, which seems to be a too-literal acceptance of the 1953 Borden-Vickers hypothesis), and then moving on until it disappeared from the scope 30 miles west of Bentwaters.

This radar-tracking of the passage of the unidentified target was matched by concurrent visual observations, by personnel on the ground looking up and also from an overhead aircraft looking down. Both visual reports involved only a light described as blurred out by its high speed; but since the aircraft (identified as a C-47 by the Lakenheath noncom) was flying at only 4,000 feet, the altitude of the unknown object is bracketed within rather narrow bounds. (No mention of any sonic boom appears; but the total number of seemingly quite credible reports of UFO's moving at speeds far above sonic values and yet not emitting booms is so large that one must count this as just one more instance of many currently inexplicable phenomena associated with the UFO problem.) The reported speed is not fast enough for a meteor, nor does the low-altitude flat trajectory and absence of a concussive shock wave match any meteoric hypothesis. That there was visual confirmation from observation points both above and below this fast-moving radar-tracked object adds still further credence to, and increases scientific interest in, the three Bentwaters radar sightings of the previous hour.

Apparently immediately after the 2255Z events, Bentwaters GCA alerted GCA Lakenheath, to the WNW. The answers to Questions 2(A) and 2(B) of the AFR200-2 format (on p. 253 of the Condon Report) *seem* to imply that Lakenheath ground observers were alerted in time to see a luminous object come in, at an estimated altitude of 2,000–2,500 feet, and on a southwest heading. The lower estimated altitude and the altered heading do not match the Bentwaters sighting, and the ambiguity inherent in the AFR200-2 format simply cannot be eliminated here, so the precise timing is not certain. At or subsequent to the Bentwaters alert message, Lakenheath ground observers saw a luminous object come in out of the northeast at low altitude, then *stop,* and take up an easterly heading and resume motion eastward out of sight.

The precise sequence of the subsequent observations is not clearly deducible from the Lakenheath TWX sent in compliance with AFR200-2, but it is clear from the report that many interesting and scientifically baffling events soon took place. No followup, from Bluebook or other USAF sources, was undertaken, and so this potentially very important case, like hundreds of others, simply went into the Bluebook files unclarified. I am forced to stress that nothing reveals so clearly the past years of scientifically inadequate UFO investigation as a few days' visit to Wright-Patterson AFB and a diligent reading of Bluebook case reports. No one with any genuine scientific interest in solving the UFO problem would have let accumulate so many years of reports like this one without seeing to it that the UFO reporting and followup investigations were brought into entirely different status from that in which they have lain for over twenty years. Deficiencies having been noted, I next catalog, without benefit of the exact time-ordering that is so crucial to full assessment of any UFO event, the intriguing observations and events at or near Lakenheath subsequent to the 2255Z alert from Bentwaters.

*Nonchronological Summary of Lakenheath Sightings, 2255Z-0330Z*

(1) *Visual observations from the ground.* As noted above, following the 2255Z alert from GCA Bentwaters, USAF ground observers at the Lakenheath RAF Station observed a luminous object come in on a southwesterly heading, stop, and then move out of sight to the east. Subsequently, at an unspecified time, two moving white lights were seen, and "ground observers stated one white light joined up with another and both disappeared in formation together" (recall earlier radar observations of merging of targets seen by Bentwaters GCA). No discernible features of these luminous sources were noted by ground observers, but both the observers and radar operators concurred in their description that "the objects [were] traveling at terrific speeds and then stopping and changing course immediately." In a passage of the original Bluebook report which was for some reason not included in the version presented in the Condon Report, this concordance of radar and visual observations is underscored: "Thus two radar sets [i.e., Lakenheath GCA and RATCC radars] and three ground observers report substantially same." Later in the original Lakenheath report, this same concordance is reiterated: "the fact that radar and ground visual observations were made on its rapid acceleration and abrupt stops certainly lend credulance (sic) to the report."

Since the date of this incident coincides with the date of peak frequency of the Perseid meteors, one might ask whether any part of the visual observations could have been due to Perseids. The basic Lakenheath report to Bluebook notes that the ground observers reported "unusual amount of shooting stars in sky," indicating that the erratically moving light(s) were readily distinguishable from meteors. The report further remarks that "the objects seen were definitely not shooting stars as there were no trails as are usual with such sightings." Furthermore, the stopping and course reversals are incompatible with any such hypothesis.

AFR200-2 stipulates that observer be asked to compare UFO to the size of various familiar objects when held at arm's length. In answer to that item, the report states: "One observer from ground stated on first observation object was about size of golf ball. As object continued in flight it became a 'pin point.' " Even allowing for the usual inaccuracies in such estimates, this further rules out Perseids, since that shower yields only meteors of quite low luminosity.

In summary of the ground-visual observations, it appears that three ground observers at Lakenheath saw at least two luminous objects, saw these over an extended although uncertain time period, saw them execute sharp course-changes, saw them remain motionless at least once, saw two objects merge into a single luminous object at one juncture, and reported motions in general accord with concurrent radar observations. These observations in themselves are scientifically interesting. Neither astronomical nor aeronautical explanations, nor any meteorological-optical explanations, match well those reported phenomena. One could certainly wish for a far more complete and time-fixed report on these visual observations, but even the above information suffices to suggest some unusual events. This impression is reinforced when the ground-radar observations from Lakenheath and the airborne-visual and airborne-radar observations made near Lakenheath are examined.

(2) *Ground-radar observations at Lakenheath.* The GCA surveillance radar at Lakenheath is identified as a CPN-4, while the RATCC search radar was a CPS-5 (as the noncom correctly recalled in his letter). Because the report makes clear that these two sets were concurrently following the unknown targets, it is relevant to note that they have different wavelengths, pulse repetition frequencies, and scan rates, facts (for reasons that need not be elaborated here) tending to rule out several radar-anomaly hypotheses (e.g., interference echoes from a distant radar, second-time-around effects, anomalous propagation). However, the reported maneuvers are so unlike any of those effects that it seems almost unnecessary to confront those possibilities here.

Again, the AFR200-2 format limitations plus the other deficiencies in reporting UFO events preclude reconstruction in detail, and in sequence, of all the relevant events. I get the impression that the first object seen visually by ground observers was not radar-tracked, although this is unclear from the report to Bluebook. One target whose motions were jointly followed both on the CPS-5 at the Radar Air Traffic Control Center and on the shorter-range, faster-scanning CPN-4 at the Lakenheath GCA unit was tracked

> from 6 miles west to about 20 miles SW where target stopped and assumed a stationary position for five minutes. Target then assumed a heading northwesterly [I presume this was intended to read "northeasterly," and the noncom so indicates in his recollective account of what appear to be the same maneuvers] into the Station and stopped two miles NW of Station. Lakenheath GCA reports three to four additional targets were doing the same maneuvers in the vicinity of the Station. Thus two radar sets and three ground observers report substantially same.

(This quotation includes the full passage omitted from the Condon Report version; note that it *seems* to imply that this devious path with two periods of stationary hovering was also reported by the visual observers. However, the latter is not entirely certain because of ambiguities in the structure of the basic report.)

At some time, which context seems to place somewhat later in the night (the radar sightings went on until about 0330Z), "Lakenheath Radar Air Traffic Control Center observed object 17 miles east of Station making sharp rectangular course of flight. This maneuver was not conducted by circular path but on right angles at speeds of 600-800 mph. Object would stop and start with amazing rapidity." The report remarks that "the controllers are experienced and technical skills were used in attempts to determine just what the objects were. When the target would stop on the scope, the MTI was used. However, the target would still appear on the scope." (The latter is puzzling. MTI, Moving Target Indication, is a standard feature on search or surveillance radars which eliminates ground returns and returns from large buildings and other motionless objects. This very curious display of stationary modes, while the MTI was on, further discredits the hypothesis of anomalous propagation of ground returns. It was as if the unidentified target, while seeming to hover motionless, was actually undergoing small-amplitude but high-speed jittering motion to yield a scope-displayed return despite the MTI. Since just such jittery motion has been reported in *visual* UFO sightings on many occasions, and since the coarse resolution of a PPI [Plan-Position Indicator] display would not permit radar-detection of such motion if its amplitude were below, say, one or two hundred meters, this could conceivably account for the persistence of the displayed return during the episodes of "stationary" hovering, despite use of MTI.)

The portion of the radar sightings just described seems to have been vividly recollected by the retired USAF noncom who first called this case to the attention of the Colorado group. Sometime after the initial Bentwaters alert, he had his men at the RATCC scanning all available scopes, various scopes set at various ranges. He wrote:

> One controller noticed a stationary target on the scopes about 20 to 25 miles southwest. This was unusual, as a stationary target should have been eliminated unless it was moving at a speed of at least 40 to 45 knots. And yet we could detect no movement at all. We watched this target on all the different scopes for several minutes and I called the GCA Unit at [Lakenheath] to see if they had this target on their scopes also. They confirmed that the target was on their scope in the same geographical location. As we watched, the stationary target started moving at a speed of 400 to 600 mph in a north-northeast direction until it reached a point about 20 miles north-northwest of [Lakenheath]. There was no slow start or build-up to this speed—it was constant from the second it started to move until it stopped.

This description, written twelve years after the event, matches the 1956 intelligence report from the Lakenheath USAF unit so well, even seeming to avoid the typographical direction-error that the Lakenheath TWX contained, that one can only assume that he was deeply impressed by this whole incident. (That, of course, is further indicated by the very fact that he wrote the Colorado group about it in the first place.) His letter (Condon Report, p. 249) adds that "the target made several changes in location, always in a straight line, always at about 600 mph and always from a standing or stationary point to his next stop at constant speed—no build-up in speed at all—these changes in location varied from 8 miles to 20 miles in length—no set pattern at any time. Time spent stationary between movements also varied from 3 or 4 minutes to 5 or 6 minutes." Because his account jibes so well with the basic Bluebook file report in the several particulars in which it can be checked, the foregoing quotation from the letter as reproduced in the Condon Report stands as meaningful indication of the highly unconventional behavior of the unknown aerial target. Even allowing for some recollective uncertainties, the noncom's description of the behavior of the unidentified radar target lies so far beyond any meteorological, astronomical, or electronic explanation as to challenge any suggestion that UFO reports are of negligible scientific interest.

The noncom's account indicates that observers plotted the discontinuous stop-and-go movements of the target for some tens of minutes before deciding to scramble RAF interceptors to investigate. That third major aspect of the Lakenheath events must now be considered. (The delay in scrambling interceptors is noteworthy in *many* Air Force–related UFO incidents of the past 20 years. I believe this reluctance stems from unwillingness to take action lest the decision-maker be accused of taking seriously a phenomenon which the Air Force officially treats as nonexistent.)

(3) *Airborne radar and visual sightings by Venom interceptor*. An RAF jet interceptor, a Venom single-seat subsonic aircraft equipped with an air-intercept (AI) nose radar, was scrambled, according to the basic Bluebook report, from Waterbeach RAF Station, which is located about 6 miles north of Cambridge and some 20 miles southwest of Lakenheath. Precise time of the scramble does not appear in the report to Bluebook, but we may infer from the noncom's account that it was near midnight. Both the noncom's letter and the contemporary intelligence report make clear that Lakenheath radar had one of their unidentified targets on-scope as the Venom came in over the Station from Waterbeach. The TWX to Bluebook states: "The aircraft flew over RAF Station Lakenheath and was vectored toward a target on radar 6 miles east of the field. Pilot advised he had a bright white light in sight and would investigate. At thirteen miles west [east?] he reported loss of target and white light."

Clearly, then, the UFO that the Venom first tried to intercept was being monitored via three distinct physical "sensing channels." It was being recorded by *ground radar*, by *airborne radar*, and *visually*. Many scientists are entirely unaware that Air Force files contain such UFO cases, for this very interesting variety of case has never been stressed in USAF discussion of its UFO records. Note, in fact, the similarity to the 1957 RB-47 case (Case 1 above) in the evidently simultaneous loss of visual and airborne-radar signal here. One wonders if ground radar also lost it simultaneously with the Venom pilot's losing it, but, as is so typical of AFR200-2 reports, incomplete reporting precludes clarification. Nothing in the Bluebook file suggests that anyone at Bluebook took any trouble to run down that point or the many other residual questions that are so evident here. The file does, however, include a lengthy dispatch from Capt. G. T. Gregory, then the Bluebook officer, a dispatch that proposes a series of what I must term wholly irrelevant hypotheses about Perseid meteors with "ionized gases in their wake which may be traced on radarscopes," and inversions that "may cause interference between two radar stations some distance apart." Such basically irrelevant remarks are all too typical of Bluebook critique over the years. The file also includes a discussion by J. A. Hynek, a Bluebook consultant, who also toys with the idea of possible radar returns from meteor wake-ionization. Not only are the radar frequencies here about two orders of magnitude too high to afford even marginal likelihood of meteor-wake returns, but there is absolutely no kinematic similarity between the reported UFO movements and the essentially straight-line hypersonic movement of a meteor, to cite just a few of the objections to meteor hypotheses. Hynek's memorandum on the case makes some suggestions about the need for upgrading Bluebook operations, and then closes with the remarks: "The Lakenheath report could constitute a source of embarrassment to the Air Force; and should the facts, as so far reported, get into the public domain, it is not necessary to point out what excellent use the several dozen UFO societies and other 'publicity artists' would make of such an incident. It is, therefore, of great importance that further information on the technical aspects of the original observations be obtained, without loss of time, from the original observers." That memo of October 17, 1956, is followed in the case file by Captain Gregory's November 26, 1956, reply, in which he concludes that "our original analyses of anomalous propagation and astronimical is more or less correct (sic)"; and there the case investigation seems to end, at the same casually closed level at which hundreds of past UFO cases have been closed out at Bluebook with essentially no real scientific critique. It is unfortunate that "the facts, as so far reported" did not get into the public domain, along with the facts on innumerable other Bluebook cases that should have long ago startled the scientific community just as they startled me when I began to study those astonishing files.

Returning to the account of the Venom pilot's attempt to make an air-intercept on the Lakenheath unidentified object, the original report goes on to note that, after the pilot lost both visual and radar signals, "RATCC vectored him to a target 10 miles east of Lakenheath and pilot advised target was on radar and he was 'locking on.' " Although here we are given no information on the important point of whether he also saw a luminous object, as he got a radar lock-on, we definitely have another instance of at least two-channel detection. The concurrent detection of a single radar target by a ground radar and an airborne radar under conditions such as these, where the target proves to be a highly maneuverable object (see below), categorically rules out any conventional explanations involving, say, large ground structures and propagation anomalies. That MTI was being used on the ground radar also excludes that, of course.

Suddenly the Venom lost radar lock-on as it neared the unknown target. RATCC reported that "as the Venom passed the target on radar, the target began a tail chase of the friendly fighter." RATCC asked the Venom pilot to acknowledge this turn of events and he did, saying "he would try to circle and get behind the target." His attempts were unsuccessful; the report to Bluebook says only, "Pilot advised he was unable to 'shake' the target off his tail and requested assistance." The noncom's letter is more detailed and emphatic. He first remarks that the UFO's sudden evasive movement into tail position was so swift that he missed it on his own scope, "but it was seen by the other controllers." His letter goes on to note that the Venom pilot "tried everything—he climbed, dived, circled, etc., but the UFO acted like it was glued right behind him, always the same distance, very close, but we always had two distinct targets." Again, note the incompleteness of the basic report. We are not told whether the pilot knew the UFO was pursuing his Venom by virtue of some tail-radar warning device of a type often used on fighters (none is alluded to), or because he could *see* a luminous object in pursuit. However, the available information does make quite clear that the pursuit was being observed on ground radar, and the noncom's recollection puts the duration of the pursuit at perhaps ten minutes before the pilot elected to return to his base. Very significantly, the intelligence report from Lakenheath to Bluebook quotes this first pilot as saying "clearest target I have ever seen on radar," which again eliminates a number of hypotheses and argues most cogently for the scientific significance of the whole episode.

The noncom recalled that, as the first Venom returned to Waterbeach aerodrome when fuel ran low, the UFO followed him a short distance and then stopped; that important detail is not in the Bluebook report. A second Venom was then scrambled, but no intercepts were accomplished in the short time before a malfunction forced it to return to Waterbeach.

*Discussion*

The Bluebook report material indicates that other radar unknowns were being observed at Lakenheath until about 0330Z. Since the first radar unknowns appeared near Bentwaters at about 2130Z on August 13, and the Lakenheath events terminated near 0330Z on August 14, the total duration of this episode was about six hours. The case includes an impressive number of scientifically provocative features:

(1) At least three separate instances occurred in which one ground-radar unit, GCA Bentwaters, tracked some unidentified target for tens of miles across its scope at speeds in excess of Mach 3. Since even today no nation has disclosed military aircraft capable of flight at such speeds (we may exclude the X-15), and since that speed is much too low to fit any meteoritic hypothesis, this first feature (entirely omitted from discussion in the Condon Report) is quite puzzling. However, Air Force UFO files and other sources contain many such instances of nearly hypersonic speeds of radar-tracked UFO's.

(2) In one instance, about a dozen low-speed (order of 100 mph) targets moved in loose formation led by three closely-spaced targets, the assemblage yielding consistent returns over a path of about 50 miles, after which they merged into a single large target, remained motionless

for some 10–15 minutes, and then moved offscope. Under the reported wind conditions, not even a highly contrived meteorological explanation invoking anomalous propagation and inversion-layer waves would account for this sequence observed at Bentwaters. The Condon Report omits all discussion of items (1) and (2), for reasons that I find difficult to understand.

(3) One of the fast-track radar sightings at Bentwaters, at 2255Z, coincided with visual observations of some very high-speed luminous source seen by both a tower operator on the ground and by a pilot aloft who saw the light moving in a blur below his aircraft at 4,000 feet. The radar-derived speed was given as 2,000–4,000 mph. Again, meteors won't fit such speeds and altitudes, and we may exclude aircraft for several evident reasons, including absence of any thundering sonic boom that would surely have been reported if any hypothetical secret 1956-vintage hypersonic device were flying over Bentwaters at less than 4,000 feet that night.

(4) Several ground observers at Lakenheath saw luminous objects exhibiting nonballistic motions, including dead stops and sharp course reversals.

(5) In one instance, two luminous white objects merged, as seen from the ground at Lakenheath. This wholly non-meteor and non-aeronautical phenomenon is actually a not-uncommon feature of UFO reports during the last two decades. For example, radar-tracked merging of two targets that veered together sharply before joining up was reported over Kincheloe AFB, Michigan, in a UFO report that also appears in the Condon Report (p. 164), where it is quite unreasonably attributed to "anomalous propagation."

(6) Two separate ground radars at Lakenheath, having rather different radar parameters, were concurrently observing movements of one or more unknown targets over an extended period of time. Seemingly stationary hovering modes were repeatedly observed, and this despite use of MTI. Seemingly "instantaneous" accelerations from rest to speeds of the order of Mach 1 were repeatedly observed. Such motions cannot be readily explained in terms of any known aircraft flying then or now, and they also fail to fit known electronic or propagation anomalies. The

Bluebook report gives the impression (somewhat ambiguously, however) that some of these two-radar observations were coincident with ground-visual observations.

(7) In at least one instance, the Bluebook report makes clear that an unidentified luminous target was seen visually from the air by the pilot of an interceptor while getting simultaneous radar returns from the unknown with his nose radar concurrent with ground-radar detection of the same unknown. This is highly significant, for it entails *three* separate detection channels all recording the unknown object.

(8) In at least one instance, there was simultaneous radar and visual disappearance of the UFO. This is akin to similar events in other known UFO cases, yet is not easily explained in terms of conventional phenomena.

(9) Attempts of the interceptor to close on one target seen both on ground radar and on the interceptor's nose radar led to a puzzling rapid interchange of roles as the unknown object moved into tail-position behind the interceptor. While under continuing radar observation from the ground, with both aircraft and unidentified object clearly displayed on the Lakenheath ground radars, the pilot of the interceptor tried unsuccessfully to break the tail chase for some minutes. No ghost-return or multiple-scatter hypothesis can explain such an event.

I believe that this sequence of baffling events, involving so many observers and so many distinct observing channels and exhibiting such unconventional features, should have led to the most intensive Air Force inquiries. But I would have to say precisely the same about dozens of other inexplicable Air Force–related UFO incidents reported to Bluebook since 1947. What the Lakenheath case shows all too well is that highly unusual events have been occurring under circumstances where any organization with even passing scientific curiosity should have responded vigorously, yet the Air Force UFO program has repeatedly exhibited just as little response as I have noted in this incident. The Air Force UFO program, contrary to the impression held by most scientists here and abroad, has been an exceedingly superficial and generally quite incompetent program. Suggestions from Air Force press offices that "the best scientific talents available to the U.S. Air Force" have been brought to bear on the UFO question are far from the truth and have misled the scientific community, here and abroad, into thinking that careful investigations were yielding solid conclusions to the effect that the UFO problem was a nonsense problem. The Air Force has given us all the impression that its UFO reports involved only misidentified phenomena of conventional sorts. That, I submit, is far from correct, and the Air Force has not responsibly discharged its obligations to the public in conveying so gross a misimpression for twenty years. Let me stress that incompetence, not conspiracy, is my charge.

The Condon Report, although disposed to suspect that some sort of anomalous radar propagation might be involved (I record here my objection that the Condon Report exhibits repeated instances of misunderstanding of the limits of anomalous propagation effects), does concede that Lakenheath is an "unexplained" case. Indeed, the report ends its discussion with the quite curious admission that, in the Lakenheath episode, "the probability that at least one genuine UFO was involved appears to be fairly high."

Whatever the meaning of the phrase "one genuine UFO," my own position is that the Lakenheath case exemplifies a disturbingly large group of UFO reports in which the apparent degree of scientific inexplicability is so great that, instead of being ignored and laughed at, those cases should all along since 1947 have been drawing the attention of a large body of the world's best scientists. Had the latter occurred, we might now have some answers, some clues to the real nature of the UFO phenomena. But twenty-two years of inadequate UFO investigations have kept this stunning scientific problem out of sight and under a very broad rug called Project Bluebook, whose final termination on December 18, 1969, ought to mark the end of an era and the start of a new one relative to the UFO problem.

More specifically, with cases like Lakenheath and the 1957 RB-47 case and many others equally puzzling that are to be found within the Condon Report, I contest Condon's principal conclusion "that further extensive study of UFO's probably cannot be justified in the expectation that science will be advanced thereby." And I contest the endorsement of such a conclusion by a panel of the National Academy of Sciences, an endorsement that appears to be based upon essentially *no* independent scientific cross-checking of case material in the report. Finally, I question the judgment of those Air Force scientific offices and agencies that have accepted so weak a report.

*The Extraterrestrial Hypothesis*

In this Lakenheath UFO episode we have evidence of some phenomena defying ready explanation in terms of present-day science and technology, some phenomena that include enough suggestion of intelligent control (like the tail-chase incident), or some broadly cybernetic equivalent thereof, that it is difficult for me to see any reasonable alternative to the hypothesis that *something in the nature of extraterrestrial devices engaged in something in the nature of surveillance lies at the heart of the UFO problem.* That is the hypothesis that my own study of the UFO problem leads me to regard as most probable in terms of my present information. This is, like all scientific hypotheses, a working hypothesis to be accepted or rejected only on the basis of continuing investigation. Present evidence surely does not amount to incontrovertible proof of the extraterrestrial hypothesis. What I find scientifically dismaying is that, while a large body of UFO evidence now seems to point in no other direction than the extraterrestrial hypothesis, the profoundly important implications of that possibility are going unconsidered by the scientific community because this entire problem has been imputed to be little more than a nonsense matter unworthy of serious scientific attention. Those overtones have been generated almost entirely by scientists and others who have done essentially no real investigation of the problem-area in which they express such strong opinions. Science is not supposed to proceed in that manner, and this AAAS symposium should see an end to such approaches to the UFO problem.

Put more briefly, doesn't a UFO case like Lakenheath warrant more than a mere shrug of the shoulders from scientists?

## Case 3. Haneda AFB, Tokyo, Japan, August 5–6, 1952

*Summary:* USAF control tower operators at Haneda AFB observed an unusually bright bluish white light to their northeast, alerted the GCI radar unit at Shiroi, which then called for a scramble of an F-94 interceptor after getting radar returns in same general area. GCI ground radar vectored the F-94 to an orbiting unknown target, which the F-94 picked up on its airborne radar. The target then accelerated out of the

F-94's radar range after 90 seconds of pursuit that was followed also on the Shiroi GCI radar.

*Introduction*

The visual and radar sightings at Haneda AFB, Japan, on August 5–6, 1952, represent an example of a long-puzzling case, diagnosed as "unidentified" by Project Bluebook and chosen for analysis in the Condon Report. In the latter, it is explained in terms of a combination of diffraction and mirage distortion of the star Capella, as far as the visual parts are concerned, while the radar portions are attributed to anomalous propagation. I find very serious difficulties with those "explanations" and regard them as typical of a number of rather casually advanced explanations of long-standing UFO cases that appear in the Condon Report. It is of particular interest to examine carefully the details of this case and then the basis of the Condon Report's explanation, as examples of how the report disposed of old "classic cases."

Haneda AFB, active during the Korean War, lay about midway between central Tokyo and central Yokohama, adjacent to Tokyo International Airport. The 1952 UFO incident began with visual sightings of a brilliant object in the northeastern sky, as seen by two control tower operators going on duty at 2330 LST (all times hereafter will be LST). It will serve brevity to introduce some code name for these men and for several officers involved, since neither the Condon Report nor my copies of the original Bluebook case file show names (excised from later copies in accordance with Bluebook practice withholding witness-names in UFO cases):

| *Coded designation* | *Identification* |
|---|---|
| Airman A | One of two Haneda tower operators who first sighted light; rank was A/3c. |
| Airman B | Second Haneda tower operator to first sight light; A/1c. |
| Lt. A | Controller on duty at Shiroi GCI unit up to 2400, August 5; 1st Lt. |
| Lt. B | Controller at Shiroi after 0000, August 6; 1st Lt. |
| Lt. P | Pilot of scrambled F-94; 1st Lt. |
| Lt. R | Radar officer in F-94; 1st Lt. |

Shiroi GCI Station, manned by the 528th AC&W (Aircraft Control and Warning) Group, lay approximately 20 miles northeast of Haneda (35°49′ N, 140°2′ E) and had a CPS-1 10-cm search radar plus a CPS-4 10-cm height-finding radar. Two other USAF facilities figure in the incident: Tachikawa AFB, just over 20 miles WNW of Haneda, and Johnson AFB, almost 30 miles NW of Haneda. The main radar incidents center over the north extremity of Tokyo Bay, roughly midway from central Tokyo to Chiba across the Bay.

The Bluebook case file on this incident contains 25 pages, and since the incident predates promulgation of AFR200-2, the strictures on time-reporting and other matters are not here so bothersome as in the Lakenheath case of 1956, discussed above. Nevertheless, the same kind of disturbing internal inconsistencies are present; in particular, times given for specific events vary in different portions of the file. One of these, stressed in the Condon Report, will be discussed explicitly below; but for the rest, I shall use those times which appear to yield the greatest overall internal consistency. This will introduce no serious errors, since the uncertainties generally involve only one or two minutes and, except for the cited instance, do not alter any important implications regardless of which cited time is used. The overall duration of the visual and radar sightings is about 50 minutes near midnight. The items of main interest occurred between 2330 and 0020, approximately.

Although this case involves both visual and radar observations of unidentified objects, careful examination does not support the view that the same object was ever assuredly seen visually and on radar at the same time, with the possible exception of the very first radar detection just after 2330. Thus it is not a "radar-visual" case, in the more significant sense of concurrent two-channel observations of an unknown object. This point will be discussed further below.

*Visual Observations*

(1) *First visual detection.* At 2330, Airmen A and B, walking across the ramp at Haneda AFB to go on the midnight shift at the airfield control tower, noticed an "exceptionally bright light" in their northeastern sky. They went immediately to the tower to alert two other on-duty controllers to it and to examine it more carefully with the aid of the 7 ×

50 binoculars available in the tower. The Bluebook file notes that controllers on duty "had not previously noticed it because the operating load had been keeping their attention elsewhere."

(2) *Independent visual detection at Tachikawa AFB.* About ten minutes later, according to the August 12, 1952, Air Intelligence Information Report (IR-35-52) in the Blue book case file, Haneda was queried about an unusually bright light by controllers at Tachikawa AFB, 21 miles to their WNW. IR-35-52 states: "The control tower at Tachikawa Air Force Base called Haneda tower at approximately 2350 to bring their attention to a brilliant white light over Tokyo Bay. The tower replied that it had been in view for some time and that it was being checked."

This feature of the report is significant in two respects: (1) It indicates that the luminous source was of sufficiently unusual brilliance to cause two separate groups of Air Force controllers at two airfields to respond independently and to take alert-actions; and (2) more significantly, the fact that the Tachikawa controllers saw the source in a direction "over Tokyo Bay" implies a line of sight distinctly south of east. From Tachikawa, even the north end of the bay lies to the ESE. Thus the intersection of the two lines of sight fell somewhere in the northern half of the bay, it would appear. As will be seen later, this is where the most significant parts of the radar tracking subsequently occurred.

(3) *Direction, intensity, and configuration of the luminous source.* IR-35-52 contains a signed statement by Airman A, a sketch of the way the luminous source looked through 7-power binoculars, and summary comments by Capt. Charles J. Malven, the FEAF intelligence officer preparing the report for transmission to Bluebook.

Airman A's statement gives the *bearing* of the source as NNE; Malven's summary specifies only NE. Presumably the witness's statement is the more reliable, and it also seems to be more precise; thus, a line-of-sight azimuth somewhere in the range of 25° to 35° east of north appears to be involved in the Haneda sightings. By contrast, the Tachikawa sighting-azimuth was in excess of 90° from north, and probably beyond 100°, considering the geography involved, a point I shall return to later.

Several different items in the report indicate the high *intensity* of the source. Airman A's signed statement refers to it as "the intense bright light over the Bay." The annotated sketch speaks of "constant brilliance across the entire area" of the (extended) source, and remarks on "the blinding effect from the brilliant light." Malven's summary even points out that "observers stated that their eyes would fatigue rapidly when they attempted to concentrate their vision on the object," and elsewhere speaks of "the brilliant blue-white light of the object." Most of these indications of brightness are omitted from the Condon Report, yet they bear on the Capella hypothesis in terms of which that report seeks to dispose of these visual sightings.

Airman A's filed statement includes the remark that "I know it wasn't a star, weather balloon or Venus, because I compared it with all three." This calls for two comments. First, Venus is referred to elsewhere in the case file, but this is certainly a matter of confusion, inasmuch as Venus had set that night before about 2000 LST. Since elsewhere in the report reference is made to Venus lying in the East, and since the only noticeable celestial object in that sector at that time would have been Jupiter, I would infer that where "Venus" is cited in the case file, one should read "Jupiter." Jupiter would have risen near 2300, almost due east, with apparent magnitude* −2.0. Thus from Airman A's assertion that the object was brighter than "Venus" we may probably infer something of the order of magnitude −3.0. From the description of the source as "blinding" or "fatiguing" to look at, I would suggest that the actual luminosity at its periods of peak value (see below) must have exceeded even magnitude −3 by a substantial margin.

Airman A's allusion to the intensity as compared with a "weather balloon" refers to the comparisons (elaborated below) with the light suspended from a pilot balloon released near the tower at 2400 that night and observed by the tower controllers to scale the size and brightness. This is a fortunate scaling comparison, because the small battery-operated lights long used in meteorological practice have a known luminosity of about 1.5 candle. Since a 1-candle source at 1 kilometer yields apparent magnitude 0.8, inverse-square scaling for the known balloon

* "Magnitude" is an inverse logarithmic measure of brightness. Jupiter, magnitude −2.0, was 100 times brighter than a medium-bright star of magnitude +3.0.

distance of 2,000 feet (see below) implies an apparent magnitude of about −0.5 for the balloon-light as viewed at time of launch. Captain Malven states, in discussing this comparison: "The balloon's light was described as extremely dim and yellow, when compared to the brilliant blue-white light of the object." Here again, I believe one can safely infer an apparent luminosity of the object well beyond Jupiter's −2.0. Thus, we have here a number of compatible indications of apparent brightness well beyond that of any star, which will later be seen to contradict explanations proposed in the Condon Report for the visual portions of the Haneda sightings.

Of further interest relative to any stellar-source hypothesis are the descriptions of the *configuration* of the object as seen with 7-power binoculars from the Haneda tower, and its approximate *angular diameter*. Fortunately, the latter seems to have been adjudged in direct comparison with an object of determinate angular size that was in view in the middle of the roughly 50-minute sighting. At 2400, a small weather balloon was released from a point at a known distance of 2,000 feet from the control tower. Its diameter at release was approximately 24 inches. (IR-35-52 refers to it as a "ceiling balloon," but the cloud-cover data contained therein are such that no ceiling balloon would have been called for. Furthermore, the specified balloon mass, 30 grams, and diameter, 2 feet, are precisely those of a standard pilot balloon for upper-wind measurement. And finally, the time [2400 LST = 1500Z] was the standard time for a pilot balloon run at that time.) A balloon of 2-foot diameter at 2,000-foot range would subtend 1 milliradian, or just over 3 minutes of arc, and this was used by the tower observers to scale the apparent angular size of the luminous source. As IR-35-52 puts it:

> Three of the operators indicated the size of the light, when closest to the tower, was approximately the same as the small ceiling balloons (30 grams, appearing 24 inches in diameter) when launched from the weather station, located at about 2000 ft from the tower. This would make the size of the central light about 50 ft in diameter, when at the 10 miles distance tracked by GCI. . . . A lighted weather balloon was launched at 2400 hours.

Thus, it would appear that an apparent angular size close to 3 minutes of arc is a reasonably reliable estimate for the light as seen by naked eye from Haneda. This is almost twice the average resolution-limit of the human eye, quite large enough to match the reported impressions that it had discernible extent, that is, was not merely a point source.

IR-35-52 gives a fairly detailed description of the object's appearance through 7-power binoculars. It is to be noted that, if the naked-eye diameter were about 3 minutes, its apparent size when viewed through 7-power binoculars would be about 20 minutes of arc, or two-thirds the naked-eye angular diameter of the full moon—quite large enough to permit recognition of the finer details cited in IR-35-52, as follows:

> The light was described as circular in shape, with brilliance appearing to be constant across the face. The light appeared to be a portion of a large round dark shape which was about four times the diameter of the light. When the object was close enough for details to be seen, a smaller, less brilliant light could be seen at the lower left-hand edge, with two or three more dim lights running in a curved line along the rest of the lower edge of the dark shape. Only the lower portion of the darker shape could be determined, due to the lighter sky which was believed to have blended with the upper side of the object. No rotation was noticed. No sound was heard.

Keeping in mind that those details are, in effect, described for an image corresponding in apparent angular size to over half a lunar diameter, the detail is by no means beyond the discernible limit. The sketch included with IR-35-52 matches the foregoing description, indicating a central disk of "constant brilliance across entire area (not due to a point source of light)," an annular dark area of overall diameter three to four times that of the central luminary, and having four distinct lights on the lower periphery, "light at lower left, small and fairly bright, other lights dimmer and possibly smaller." Finally, supportive comment is contained in the signed statement of Airman A: "After we got in the tower I started looking at it with binoculars, which made the object much clearer. Around the bright white light in the middle, there was a darker object which stood out against the sky, having little white lights along the outer edge, and a glare around the whole thing."

All of these configurational details, like the indications of a quite unstarlike brilliance, will be seen below to be almost entirely unexplainable on the Capella hypothesis. Further questions arise from examination of reported motions of the luminous source.

(4) *Reported descriptions of apparent motions of the luminous source.* Here we meet the single most important ambiguity in the Haneda case file, though the weight of the evidence indicates that the luminous object exhibited definite movement. The ambiguity arises chiefly from the way Captain Malven summarized the matter in his IR-35-52 report a week after the incident:

> The object faded twice to the East, then returned. Observers were uncertain whether disappearance was due to a dimming of the lights, rotation of object, or to the object moving away at terrific speed, since at times of fading the object was difficult to follow closely, except as a small light. Observers did agree that when close, the object did appear to move horizontally, varying apparent position and speed slightly.

Aside from the closing comment, all of Malven's summary remarks could be interpreted as implying either solely radial motion (improbable because it implies that the Haneda observers just happened to be in precisely the spot from which no crosswise velocity component could be perceived) or else merely illusion of approach and recession due to some intrinsic or extrinsic time-variation in apparent brightness.

Airman A, in contrast, seems to refer to distinct motions, including transverse components: "I watched it disappear twice through the glasses. It seemed to travel to the East and gaining altitude at a very fast speed, much faster than any jet. Every time it disappeared it returned again, except for the last time when the jets were around. It seemed to know they were there. As for an estimate of the size of the object—I couldn't even guess." Recalling that elsewhere in that same signed statement this tower controller had given the observed direction of the object as NNE, his specification that the object "seemed to travel to the East" seems quite clearly to imply a nonradial motion, since, if only an impression of the latter were involved, one would presume he would have spoken of it in some such terms as "climbing out rapidly to the NNE." Since greater weight is presumably to be placed on direct-witness testimony than on another's summary thereof, it appears necessary to assume that not mere radial recession but also transverse components of recession, upward and toward the east, were observed.

That the luminous source varied substantially in angular size is made very clear at several points in the file: One passage already cited discusses the "size of the light, when closest to the tower," while, by contrast, another says that: "At the greatest distance, the size of the light appeared slightly larger than Venus [Jupiter], approximately due East of Haneda, and slightly brighter." Jupiter was then near quadrature with angular diameter of around 40 seconds of arc. Since the naked eye is a poor judge of comparative angular diameters that far below the resolution limit, little more can safely be read into that statement than the conclusion that the object's luminous disk diminished quite noticeably and its apparent brightness fell to a level comparable to or a bit greater than Jupiter's when at greatest perceived distance. This is another basis for concluding that when at *peak* brilliance it must have been considerably brighter than Jupiter's −2.0, a conclusion already reached by other arguments above.

In addition to exhibiting apparent recession, eastward motion, and climb to disappearance, the source also disappeared for at least one other period far too long to be attributed to any scintillation or other such meteorological optical effect: "When we were about half way across the ramp," Airman A stated, "it disappeared for the first time and returned to approximately the same spot about 15 seconds later." There were scattered clouds over Haneda at around 15–16,000 feet, and a very few isolated clouds lower down, yet it was full moon that night, and, if patches of clouds had drifted very near the controllers' line of sight to the object, they could be expected to have seen the clouds. (The upper deck was evidently thin, for Malven notes in his report, "The F-94 crew reported exceptional visibility and stated that the upper cloud layer did not appreciably affect the brilliancy of the moonlight.") A thin cloud interposed between observer and a distant luminous source would yield an impression of dimming and *enhanced* effective angular diameter, not dimming and *reduced* apparent size, as reported here. I believe the described "disappearances" cannot, in view of these several considerations, reasonably be attributed to cloud effects.

These, then, are the essential features of the Haneda report dealing with the visual observations of some bright luminous source that initiated the alert and that led to the ground- and airborne-radar observations yet to be described. Before turning to those—the more significant portion of the sighting—it will be best to examine the Bluebook and the Condon Report attempts to explain the visual observations.

*Bluebook Critique of the Visual Sightings*

In IR-35-52, Captain Malven offers just one hypothesis, and that only in passing: He speculates briefly on whether "reflections off the water [of the bay, I presume] were . . . sufficient to form secondary reflections off the lower clouds," and by the latter he means "isolated patches of thin clouds reported by the F-94 crew as being at approximately 4000 feet." He adds that "these clouds were not reported to be visible by the control tower personnel," which, in view of the 60-mile visibility cited elsewhere in the case file and in view of the full moon then near the local meridian, suggests that those lower clouds must have been exceedingly widely scattered to escape detection by the controllers.

What Malven seems to offer there, as a hypothesis for the observed visual source, is cloud-reflection of moonlight, and in a manner all too typical of many other curious physical explanations one finds scattered through Bluebook files, he reveals a lack of appreciation of what is central to the issue. If he wants to talk about cloud-reflected moonlight, why render a poor argument even weaker by invoking not direct moonlight but moonlight secondarily reflected off the surface of Tokyo Bay? Without even considering further that odd twist in his tentative hypothesis, it is sufficient to note that even direct moonlight striking a patch of cloud is not "reflected" in any ordinary sense of that term. It is scattered from the cloud droplets and thereby serves not to create any image of a discrete light source of blinding intensity that fatigues observers' eyes and does the other things reported by the Haneda observers, but rather serves merely to palely illuminate a passing patch of cloud material. A very poor hypothesis.

Malven uses good judgment in not stressing that hypothesis. He does add that there was some thunderstorm activity reported that night to the northwest of Tokyo, but mentions that there was no reported electrical activity therein. Since the direction is opposite to the line of sight and since the reported visual phenomena bear no relation to lightning effects, the report carries that point no further.

Finally, Malven mentions very casually an idea that I have encountered repeatedly in Bluebook files and nowhere else in my studies of atmospheric physics, namely, "reflections off ionized portions of the atmosphere." He states: "Although many sightings might be attributed to visual and electrical reflections off ionized areas in the atmosphere, the near-perfect visibility on the night of the sighting, together with the circular orbit of the object, would tend to disprove this theory." Evidently he rejects the "ionized areas" hypothesis on the ground that presence of such areas is probably ruled out by the unusually good visibility. I trust that, for most readers of this discussion, I would only be belaboring the obvious to remark that Bluebook mythology about radar and visual "reflections" off "ionized regions" in the clear atmosphere (which mythology I have recently managed to trace back even to pre-1950 Air Force documents on UFO reports) has no known basis in fact.

Although the final Bluebook evaluation of this entire case is "unidentified," indicating that none of the above was regarded as an adequate explanation of even the visual features of the report, one cannot overlook extremely serious deficiencies in the basic reporting and the interrogation and follow-up here. This incident occurred in that period which my own studies lead me to describe as sort of a highwater mark for Project Bluebook. Capt. Edward J. Ruppelt was then Bluebook Officer at Wright-Patterson AFB, and both he and his superiors were treating the UFO problem more seriously than the USAF had done at any other time in the entire twenty-two years of the project. Neither before nor after 1952–53 were there as many efforts made to assemble case information, to go out and actually check in the field on sightings, and so forth. Yet it should be all too apparent, even at this point in the discussion of the Haneda case, that quite basic points were not run to ground and pinned down. In his *Report on the Unidentified Flying Objects* (New York: Ace Books, 1956) Ruppelt speaks of this case as though it were one of the most completely reported cases they had received as of mid-1952. He mentions, for example, that his office queried FEAF offices about a few points of confusion and that the replies came back with impressive promptness. If one needed some specific instance of the regrettably low scientific level of the operation of Bluebook even during this period of comparatively energetic investigation, one can find it in study of the Haneda report. Even so simple a matter as checking whether Venus was actually in the east was obviously left undone. That, I stress, is what any scientist who studies the Bluebook files as I have done will find through all the years of Air Force handling of the UFO problem: incompetence and superficiality—even at the 1952 highwater mark under Ruppelt's relatively vigorous direction.

*Condon Report Critique of the Visual Sightings*

On p. 126 of the Condon Report, the luminous source discussed above is explained as a diffracted image of the star Capella: "The most likely light source to have produced the visual object is the star Capella (magnitude 0.2), which was 8° above horizon at 37° azimuth at 2400 LST. The precise nature of the optical propagation mechanism that would have produced such a strangely diffracted image as reported by the Haneda AFB observers must remain conjectural."

Suggesting that perhaps "a sharp temperature inversion may have existed at the top of [an inferred] moist layer, below which patches of fog or mist could collect," the report continues:

> The observed diffraction pattern could have been produced by either (1) interference effects associated with propagation within and near the top of an inversion, or (2) a corona with a dark aureole produced by a mist of droplets of water of about 0.2 mm diameter spaced at regular intervals as described by Minnaert. . . . In either event, the phenomenon must be quite rare. The brightness of the image may have been due in part to "Raman brightening" of an image seen through an inversion layer.

In the final paragraph discussing this case, the Condon Report concludes that "the most probable" cause of the visual sighting is "an optical effect on a bright light source." There are some very serious difficulties with the more specific parts of the suggested explanation, and the vagueness of the other parts is sufficiently self-evident to need little comment.

First, nothing in the literature of meteorological optics discusses any diffraction-produced corona with a dark annular space extending out to three or four diameters of the central luminary, such as is postulated in the Condon Report explanation. The radial intensity pattern of a corona may be roughly described as a damped oscillatory radial variation of luminosity, with zero-intensity minima (for the simple case of a monochromatic luminary) at roughly equal intervals, and no broad light-free annulus comparable to that described in detail by the Haneda controllers. Thus, a lack of understanding of the nature of coronae is inherent in this explanation.

Second, droplets certainly do not have to be "spaced at regular intervals" to yield a corona, and Minnaert's book, *The Nature of Light and Colour in the Open Air* (New York: Dover, 1954), makes no such suggestion, another measure of misunderstanding of the meteorological optics concerned here. Nor is there any physical mechanism operating in clouds capable of yielding any such regular droplet spacing. Both Minnaert and cloud physics are misunderstood in that passage.

Third, one quickly finds, by some trial calculations, using the familiar optical relation (Exner equation) for the radial positions of the minima of the classical corona pattern, that the cited drop diameter of 0.2 mm (= 200 microns) was obtained in the Condon Report by back-calculating from a tacit requirement that the first-order minimum lay close to 3 milliradians (10 minutes of arc), for these are the values that satisfy the Exner equation for wavelength about 0.5 microns (visible light). This discloses even more thorough misunderstanding of corona optics, for that first-order minimum marks not some outer edge of a broad dark annulus as described and sketched by the Haneda tower operators, but the outer edge of the innermost annulus of *high* intensity of diffracted light.

Fourth, the computation cited yielded a droplet diameter of 200 microns, which is so large as to be found only in drizzling or raining clouds and never in thin, scattered clouds of the sort reported, clouds that scarcely attenuated the full moon's light. That is, the suggestion that "patches of fog or mist" collected under a hypothesized inversion could grow droplets of that large size is meteorologically out of the question. *If* isolated patches of clouds interposed themselves on an observer's line of light to some distant luminary, under conditions of the sort prevailing at Haneda that night, drop diameters down in the range of 10–20 microns would be the largest one could expect, and the corona size would be some 10 to 20 times greater than the 3 milliradians which was plugged into the Exner equation in the above-cited computation.

And this would, of course, not even begin to match anything observed that night.

Fifth, the vague suggestion that "Raman brightening" or other "interference effects associated with propagation within and near the top of an inversion" is involved here makes the same serious error that is made in attempted optical explanations of other cases in the Condon Report. Here we are asked to consider that light from Capella, whose altitude was about 8° above the northeastern horizon (a value that I confirm) near the time of the Haneda observations, was subjected to Raman brightening or its equivalent; yet one of the strict requirements of all such interference effects is that the ray paths impinge on the inversion surface at grazing angles of incidence of only a small fraction of a degree. No ground observer viewing Capella at 8° elevation angle could *possibly* see anything like Raman brightening, for the pertinent angular limits would be exceeded by *one or two orders of magnitude.* Added to this measure of misunderstanding of the optics of such interference phenomena in this attempted explanation is the further difficulty that, for any such situation as is hypothesized in the Condon Report explanation, the observer's eye must be located at or directly under the index-discontinuity, which would here mean up in the air at the altitude of the hypothesized inversion. But all of the Haneda observations were made from the ground level. Negation of Raman brightening leaves one more serious gap in the Capella hypothesis, since its magnitude of 0.2 lies at a brightness level well below that of Jupiter, yet the Haneda observers seem to have been comparing the object's luminosity to Jupiter's and finding it far brighter, not dimmer.

Sixth, the Condon Report mentions the independent sighting from Tachikawa AFB, but fails to bring out that the line of sight from that observing site (luminary described as lying over Tokyo Bay, as seen from Tachikawa) pointed more than 45° away from Capella, a circumstance fatal to fitting the Capella hypothesis to both sightings. Jupiter lay due east, not "over Tokyo Bay" from Tachikawa, and it had been rising in the eastern sky for many days, so it is, in any event, unlikely to have suddenly triggered an independent response at Tachikawa that night. And, conversely, the area intersection of the reported lines of sight from Haneda and Tachikawa falls in just the northern Bay area where Shiroi GCI first got radar returns and where all the subsequent radar activity was localized.

Seventh, nothing in the proffered explanations in the Condon Report confronts the reported movements and disappearances of the luminous object that are described in the Bluebook case file on Haneda. If, for the several reasons offered above, we conclude that not only apparent radial motions but also lateral and climbing motions were observed, neither diffraction nor Raman effects can conceivably fit them.

Eighth, the overall configuration as seen through 7-power binoculars, particularly with four smaller lights perceived on the lower edge of the broad, dark annulus, is not in any sense explained by the ideas qualitatively advanced in the Condon Report.

Ninth, the Condon Report puts emphasis on the point that, whereas Haneda and Tachikawa observers saw the light, airmen at the Shiroi GCI site went outside and looked in vain for it when the plotted radar position showed one or more targets to their south or south-southeast. This is correct. But we are quite familiar with both highly directional and hemi-directional light sources on our own technological devices, so the failure to detect a light from the Shiroi side does not very greatly strengthen the hypothesis that Capella was the luminary in the Haneda visual sightings. The same can be said for lack of visual observations from the F-94, which got only radar returns as it closed on its target.

I believe that it is necessary to conclude that the "explanation" proposed in the Condon Report for the visual portions of the Haneda case is almost wholly unacceptable, as are many others. We were supposed to get in the Condon Report a level of critique distinctly better than that which had come from Bluebook for many years; but much of the critique in that report is little less tendentious and ill based than that which is so dismaying in twenty-two years of Air Force discussions of UFO cases. The above stands as only one illustration of the point I make there; many more could be cited.

*Radar Observations*

Shortly after the initial visual sighting at Haneda, the tower controllers alerted the Shiroi GCI radar unit (located about 15 miles northeast of central Tokyo), asking them to look for a target somewhere northeast of Haneda at an altitude which they estimated (obviously on weak grounds) to be somewhere between 1,500 and 5,000 feet; both these figures appear in the Bluebook case file. A CPS-1 search radar and a CPS-4 height-finder radar were available at Shiroi, but only the CPS-1 picked up the target, ground clutter interference precluding useful CPS-4 returns. The CPS-1 radar was a 10-cm, 2-beam set with peak power of 1 megawatt, PRF of 400/sec, antenna tilt 3°, and scan-rate operated that night at 4 rpm. I find no indication that it was equipped with MTI, but this point is not certain.

The following list may help to put the sequence of events in a clear time order. In some instances a one- to two-minute range of time is given because the case file contains more than a single time for that event as described in separate sections of the report. I indicate 0015-16 LST as the time of first airborne radar contact by the F-94, and discuss that matter in more detail later, since the Condon Report suggests a quite different time.

| *Time (LST)* | *Event* |
|---|---|
| 2330 | Tower controllers at Haneda see bright light to NE, call Shiroi GCI within a few minutes. |
| 2330–45 | Lt. A, Shiroi radar controller on evening watch, looks for returns, finds 3-4 stationary blips to NE of Haneda on low beam of CPS-1. |
| 2345 | Lt. B comes on duty for midwatch at Shiroi; he and Lt. A discuss possible interceptor scramble. |
| 2355 | Lt. A calls Johnson AFB, asks for F-94 scramble. Fuel system trouble causes delay of 5–10 min. |
| 0001 | Lt. B has unknown in right orbit at varying speeds over north Tokyo Bay, 8 miles NE of Haneda. Loses contact again. |
| 0003–04 | F-94 airborne out of Johnson AFB, Lt. P as pilot, Lt. R, radarman. |
| 0009–10 | Shiroi alerts F-94 to airborne target to its starboard as it heads down Tokyo Bay, and Lt. P visually identifies target as C-54 in pattern to land at Haneda. Lt. B instructs Lt. P to begin search over north Bay area at flight altitude of 5000 ft. |

| *Time (LST)* | *Event* |
|---|---|
| 0012 | Shiroi regains CPS-1 contact on unknown target in right orbit over same general area seen before, target splits into three separate targets, and Lt. B vectors F-94 toward strongest of three returns. |
| 0015–16 | F-94 gets airborne radar contact on moving target at range and bearing close to vector information, has to do hard starboard turn to keep onscope as target moves with acceleration across scope. |
| 0017–18 | After 90 seconds pursuit, with no lock-on achieved, target moves offscope at high speed; Shiroi GCI tracks both unknown and F-94 into its ground clutter, where both are then lost in clutter. |
| 0033 | Shiroi releases F-94 from scramble-search. |
| 0040 | F-94 visually spots another C-54, over Johnson. |
| 0120 | F-94 lands back at Johnson. |

Thus the period 2330 on 8/5 through about 0018 on 8/6 is of present interest.

(1) *Initial attempts at radar detection from Shiroi GCI.* When, at about 2335 or so, Haneda requested Shiroi to search the area of the bay to the NE of Haneda (SSW from Shiroi, roughly), Lt. A, then duty controller at Shiroi, found his CPS-4 giving too much ground clutter to be useful for the relatively low estimated heights Haneda had suggested. Those heights are indicated as 1,500-2,000 feet in one portion of the case file, though Airman A elsewhere gave 5,000 feet as his impression of the height. Clearly, lack of knowledge of size and slant ranges precluded any exact estimates from Haneda, but they offered the above indicated impressions.

Trying both low and high beams on the CPS-1 search radar, Lt. A did detect three or four blips "at a position 050° bearing from Haneda, as reported by the tower, but no definite movement could be ascertained." The report gives no information on the range from Shiroi, nor on the inferred altitude of those several blips, and these are only the first of a substantial number of missing items of essential information that were not followed up in any Bluebook inquiries, as far as the file shows. No indication of the spacing of the several targets is given either, so it is difficult to decide whether to consider the above as an instance of "radar-visual" concurrency or not. One summary discussion in the Bluebook file so construes it: "The radar was directed onto the target by visual observations from the tower. So it can safely be assumed that both visual and radar contacts involved the same object." By contrast, the Condon Report takes the position that there were no radar observations that ever matched the visual observations. The latter view seems more justified, although the issue is basically unresolvable. Our visual target will not, in any event, match three or four radar targets, unless we can say that later on the main radar target split up into three separate radar targets, and assume that, at 2335, three or four unknown objects were airborne and motionless, with only one of these luminous and visually detectable from Haneda. That is conceivable but involves too strained assumptions to take very seriously; so I conclude that, even in this opening radar search, there was not obvious correspondence between visual and radar unknowns. As we shall see, later on there was definitely not correspondence, and also the F-94 crew never spotted a visual target. Thus, Haneda cannot be viewed as a case involving the kind of "radar-visual" concurrency which does characterize many other important cases. Nonetheless, both the visual and the radar features, considered separately, are sufficiently unusual to support the view that inexplicable events were seen and tracked there that night.

One may ask why a radar-detected object was not seen visually, and why a luminous object was not detected on search radar; and there is no fully satisfactory answer for either question. It can only be noted that there are many other such cases in Bluebook files and that these questions are part of the substantial scientific puzzle that centers around the UFO phenomena. We know that light sources can be turned off, and we do know that ECM techniques can fool radars to a certain extent. Thus, we might do well to maintain open minds when we come to such questions in UFO case analyses.

(2) *F-94 scramble.* When Lt. B came on duty at 2345, he was soon able, according to Captain Malven's summary in IR-35-52, "to make radar contact on the 50-mile high beam," whereupon he and Lt. A con-

tacted the ADCC flight controller at Johnson AFB 35 miles to their west, requesting that an interceptor be scrambled to investigate the source of the visual and the radar sightings.

An F-94B of the 339th Fighter-Interceptor Squadron, piloted by Lt. P, with Lt. R operating the APG-33 air-intercept radar, was scrambled, though a delay of over ten minutes intervened because of fuel-system difficulties during engine runup. The records show the F-94 airborne at about 0003-04, and it then took about 10 minutes to reach the Tokyo Bay area. The APG-33 set was a 3-cm (X-band) set with 50 KW power, and lock-on range of about 2,500 yards, according to my information. The system had a B-scope; i.e., it displayed target range vs. azimuth. The case file notes: "The APG-33 radar is checked before and after every mission and appeared to be working normally."

At 0009, Shiroi picked up a moving target near Haneda and alerted the F-94 crew, who had no difficulty identifying it visually as an Air Force C-54 in the Haneda pattern. The crew is quoted as reporting "exceptional visibility." Shiroi instructed the F-94 to begin searching at 5,000 feet as it got out over the Bay. But before proceeding with events of that search, a GCI detection of a moving target at about 0001 must be reviewed.

(3) *First GCI detection of orbiting object.* Just before the F-94 became airborne out of Johnson AFB, Lt. B picked up the first definitely unusual moving target, at about 0000-01. His statement in the Bluebook file reads: "At the time of the scramble, I had what was believed to be the object in radar contact. The radar sighting indicated the object to be due south of this station over Tokyo Bay and approximately eight (8) miles northeast of Haneda. The target was in a right orbit moving at varying speeds. It was impossible to estimate speed due to the short distance and times involved." That passage is quoted in the Condon Report, but not the next, which comes from Malven's summary and indicates that Lt. B only meant that it was impossible to estimate the target's speed with much accuracy. The omitted passage is interesting, for it is one of a number of indications that anomalous propagation (which is the Condon Report's explanation for the radar sightings) is scarcely creditable:

An F-94 was scrambled to investigate. The object at this time had left the ground clutter and could be tracked (on the CPS-1) at varying speeds in a right orbit. Although impossible to accurately estimate speed, Lt. B gave a rough estimate of 100-150 knots, stopping, and hovering occasionally, and a maximum speed during the second orbit (just before F-94 was vectored in) of possibly 250-300 knots.

A map accompanying IR-35-52 shows the plotted orbiting path of the unknown target. The orbit radius is approximately 4 miles, centered just off the coast from the city of Funabashi, east of Tokyo. The orbiting path is about half over land, half over water. The map-sketch, plus the file comments, imply that GCI had good contacts with the target only while it was moving out over the bay. The ground-clutter pattern of the CPS-1 is plotted on the same map (and on other maps in the file), and it seems clear that the difficulty in tracking the target through the land portion of the roughly circular orbit was that most of that portion lay within the clutter area. Presumably this set did not have MTI, which is unfortunate.

The circumference of the orbit of about 4-mile radius would be about 25 miles. Taking Lt. B's rough estimate of 100-150 knots in the first of the two circuits of this orbit (i.e., the one he detected at about 0001), a total circuit time of perhaps 12-13 minutes is indicated. Although this estimate is quite rough, it matches reasonably well the fact that it was about 0012 when it had come around again, split up into three targets, and looped onshore again with the F-94 in pursuit this time.

If the object executing the orbits had been the luminous object being watched from Haneda, it would have swung back and forth across their sky through an azimuth range of about 30°. Since no such motion seems to have been noted by the Haneda observers, I believe it must be concluded that the source they watched was distinct from the one radar-tracked in orbit.

(4) *Second orbit and F-94 intercept attempt.* The times given in Lt. B's account of this phase of the sighting do not match those given by the pilot and radarman of the F-94 in their signed statements in the file. Other accounts in the file match those of the aircrew, but not the times in Lt. B's summary. This discrepancy (about ten to twelve minutes) is

specifically noted in Malven's summary: "The ten minute difference in time between the statement by Lt. B, 528th AC&W SQ, and that reported by other personnel concerned, is believed to be a typographical error, since the statement agrees on every other portion of the sighting." That Lt. B and the aircrew were describing one and the same intercept seems beyond any doubt; and in view of Malven's quoted comment, I here use the times recorded by the aircrew and accepted as the correct times in other parts of the case file. Further comment on this will be given below.

After completing the first of the two orbits partially tracked by GCI Shiroi, the target came around again where it was out of the CPS-1 ground-clutter pattern, and Lt. B regained contact. Malven's summary comments on the next developments as follows: "At 0012 the object reportedly broke into three smaller contacts, maintaining an interval of about ¼ mile, with one contact remaining somewhat brighter. The F-94 was vectored on this object, reporting weak contact at 0015 and loss of contact at 0018. Within a few seconds, both the F-94 and the object entered the ground clutter and were not seen again."

The same portion of the incident is summarized in Lt. B's account (with different times), with the F-94 referred to by its code name "Sun Dial 20." Immediately following the part of his account referring to the first starboard orbit in which he had plotted the target's movements, at around 0001, comes the following section:

> Sun Dial 20 was ordered to search the Tokyo Bay area keeping a sharp lookout for any unusual occurrences. The object was again sighted by radar at 0017 on a starboard orbit in the same area as before. Sun Dial 20 was vectored to the target. He reported contact at 0025 and reported losing contact at 0028. Sun Dial 20 followed the target into our radar ground clutter area and we were unable to give Sun Dial 20 further assistance in re-establishing contact. Sun Dial 20 again resumed his visual search of the area until 0034, reporting negative visual sighting on this object at any time.

If Malven's suggestion of typographical error is correct, the in-contact times in the foregoing should read 0015 and 0018, and presumably 0017 should be 0012. But regardless of the precise times, the important point is that Lt. B vectored the F-94 into the target, contact was thereby achieved, and Lt. B followed the target and pursuing F-94 northeastward into his ground clutter. I stress this because, in the Condon Report, the matter of the different times quoted is offered as the sole basis for the conclusion that ground radar and airborne radar were never following the same target. This is so clearly inconsistent with the actual contents of the case file that it is difficult to understand the report rationale.

Even more certain indication that the GCI radar was tracking target and F-94 in this crucial phase is given in the accounts prepared and signed by the pilot and his radarman. The code name "Hi-Jinx" refers to the Shiroi GCI used in the air-to-ground radio transmissions that night. The F-94 pilot, Lt. P states:

> The object was reported to be in the Tokyo Bay area in an orbit to the starboard at an estimated altitude of 5,000 feet. I observed nothing of an unusual nature in this area; however, at 0016 when vectored by Hi-Jinx on a heading of 320 degrees, and directed to look for a bogie at 1100 o'clock, 4 miles, Lt. R made radar contact at 10 degrees port, 6,000 yards. The point moved rapidly from port to starboard and disappeared from the scope. I had no visual contact with the target.

And the signed statement from the radarman, Lt. R, is equally definite about these events:

> At 0015 Hi-Jinx gave us a vector of 320 degrees. Hi-Jinx had a definite radar echo and gave us the vector to intercept the unidentified target. Hi-Jinx estimated the target to be at 11 o'clock to us at a range of 4 miles. At 0016 I picked up the radar contact at 10 degrees port, 10 degrees below at 6,000 yards. The target was rapidly moving from port to starboard and a "lock on" could not be accomplished. A turn to the starboard was instigated to intercept target which disappeared on scope in approximately 90 seconds. No visual contact was made with the unidentified target. We continued our search over Tokyo Bay under Hi-Jinx control. At 0033 Hi-Jinx released us from scrambled mission.

Of particular importance here is the very close agreement of the vectoring instructions given by Shiroi GCI to the F-94 and the actual relative position at which they accomplished radar contact; GCI said 4 miles' range at the aircraft's 11 o'clock position, and they actually got radar contact with the moving target at a 6,000-yard range, 10 degrees to their port. *Nearly exact agreement,* incontrovertibly demonstrating that ground radar and airborne radar were then looking at the same moving unknown target, despite the contrary suggestions made in the Condon Report. Had the report presented all of the information in the case file, it would have been difficult for it to adhere to its curious position.

That the moving target, as seen by both ground and airborne radar, was a distinct target, though exhibiting a radar cross-section somewhat smaller than most aircraft, is spelled out in Malven's IR-35-52 summary: "Lt. B, GCI Controller at the Shiroi GCI site, has had considerable experience under all conditions and thoroughly understands the capabilities of the CPS-1 radar. His statement was that the object was a bonafide moving target, though somewhat weaker than that normally obtained from a single jet fighter." And, with reference to the airborne radar contact, the same report states: "Lt. R, F-94 radar operator, has had about seven years' experience with airborne radar equipment. He states that the object was a bonafide target, and that to his knowledge, there was nothing within an area of 15-20 miles that could give the radar echo." It is exceedingly difficult to follow the Condon Report in attributing such targets to anomalous propagation.

Not only were there no visual sightings of the orbiting target as viewed from the F-94, but neither were there any from the Shiroi site, though Lt. B specifically sent men out to watch as these events transpired. Also, as mentioned earlier, it seems out of the question to equate any of the Haneda visual observations to the phase of the incident just discussed. Had there been a bright light on the unknown object during the time it was in starboard orbit, the Haneda observers would almost certainly have reported those movements. To be sure, the case file is incomplete in not indicating how closely the Haneda observers were kept in touch as the GCI-directed radar-intercept was being carried out. But at least it is clear that the Haneda tower controllers did not describe motions of the intensely bright light that would fit the roughly circular starboard orbits of radius near four miles. Thus, we seem forced to conclude either that the target the F-94 pursued was a different one from that observed at Haneda (likely interpretation), or that it was nonluminous during that intercept (unlikely alternative, since Haneda observations did not have so large a period of nonvisibility of the source they had under observation between 2330 and 0020).

*Condon Report Critique of the Radar Sightings*

The Bluebook case file contains almost no discussion of the radar events, no suggestion of explanations in terms of any electronic or propagational anomalies. The case was simply put in the "unexplained" category back in 1952 and has remained there.

By contrast, the Condon Report regards the radar events as attributable to "anomalous propagation." Four reasons are offered (p. 126) in support of that conclusion: the tendency for targets to disappear and reappear; the tendency for the target to break up into smaller targets; the apparent lack of correlation between the targets seen on the GCI and airborne radars; and the radar invisibility of the target when visibility was "exceptionally good." Each of these four points will now be considered.

First, the "tendency for the targets to disappear and reappear" was primarily a matter of the orbiting target's moving into and out of the ground-clutter pattern of the CPS-1, as is clearly shown in the map that constitutes Enclosure #5 in the IR-35-52 report, which was at the disposal of the Colorado staff concerned with this case. Ground returns from AP (anomalous propagation) may fade in and out as ducting intensities vary, but here we have the case of a moving target disappearing into and emerging from ground clutter, while executing a roughly circular orbit some four miles in radius. I believe it is safe to assert that nothing in the annals of anomalous propagation matches such behavior. Nor could the Borden-Vickers hypothesis * of "reflections" off moving waves on inversions fit this situation, since such waves would not propagate in orbits, but would, at best, advance with the direction and speed of the mean wind at the inversion. Furthermore, the indicated target speed in the final phases of the attempted intercept was greater than that of the F-94, i.e., over 400 knots, far above wind speeds prevailing that night, so this could not in any event be squared with the (highly

* See Condon Report, pp. 136–137.

doubtful) Borden-Vickers hypothesis that was advanced years ago to account for the 1952 Washington National Airport UFO incidents.

Second, the breakup of the orbiting target into three separate targets cannot fairly be referred to as a "tendency for the target to break up into smaller targets." That breakup occurred just once, and the GCI controller chose to vector the F-94 onto the strongest of the resultant three targets. And when the F-94 initiated radar search in the *specific* area (11 o'clock at 4 miles) where that target was then moving, it immediately achieved radar contact. For the Condon Report to gloss over such definite features of the report and merely allude to all of this in language faintly suggestive of AP is objectionable.

Third, to build a claim that there was "Apparent lack of correlation between the targets seen on the GCI and airborne radars" on the sole basis of the mismatch of times listed by Lt. B and the aircrew, to ignore the specific statement by the intelligence officer filing IR-35-52 that this was a typographical error on the part of Lt. B, and, above all, to ignore the obviously close correspondence between GCI and airborne radar targeting that led to the successful radar-intercept, and finally to ignore Lt. B's statement that the F-94 "followed the target into our radar ground clutter," all amount to a highly slanted assessment of case details, details not openly set out for the reader of the Condon Report to evaluate for himself. I believe that all of the material I have here extracted from the Haneda case file fully contradicts the third of the Condon Report's four reasons for attributing the radar events to AP. I would suggest that it is precisely the impressive correlation between GCI and F-94 radar targeting on this nonvisible, fast-moving object that constitutes the most important feature of the whole case.

Fourth, the Report seems to suspect AP because of "the radar invisibility of the target when visibility was 'exceptionally good.' " This is simply unclear. The exceptional visibility in the atmosphere that night is not physically related to "radar invisibility" in any way, and I suspect this was intended to read "the invisibility of the radar target when visibility was exceptionally good." As cited above, neither the Shiroi crew nor the F-94 crew ever saw any visible object to match their respective radar targets. Under some circumstances, such a situation would indeed be diagnostic of AP. But not here, where the radar target is moving at high speed around an orbit many miles in diameter, occasionally hovering motionless (see Malven's account cited earlier), and changing speed from 100-150 knots up to 250-300 knots, and finally accelerating to well above an F-94's 375-knot speed.

Thus, *all four* of the arguments for AP offered in the Condon Report must be rejected. Those arguments seem to me to be built up by a highly selective extraction of details from the Bluebook file, by ignoring the limits of the kind of effects one can expect from AP, and by using wording that so distorts key events in the incident as to give a vague impression where the facts of the case are really quite specific.

It has, of course, taken more space to clarify this Haneda case than the case is given in the Condon Report itself. Unfortunately, this would also prove true of the clarification of some fifteen to twenty other UFO cases whose "explanation" in the Condon Report contains, in my opinion, equally objectionable features, equally casual glossing-over of physical principles, of important quantitative points. Equally serious omission of basic case information marks many of the case discussions. I have used Haneda only as an illustration of those points; but I stress that it is by no means unique. The Condon Report confronted a disappointingly small sample of the old "classic" cases, the long-puzzling cases that have kept the UFO question alive over the years, and those few that it did confront it explained away by argumentation as unconvincing as that which disposes of the Haneda AFB events in terms of diffraction of Capella and anomalous propagation. Scientifically weak argumentation is found in a large fraction of the case analyses of the Condon Report, and stands as the principal reason why its conclusions ought to be rejected.

### Case 4. Kirtland AFB, New Mexico, November 4, 1957

*Summary:* Two CAA control operators observe a lighted egg-shaped object descend to and cross obliquely the runway area at Kirtland AFB (Albuquerque), hover near the ground for tens of seconds, then climb at unprecedented speed into the overcast. On radar, it was then followed south some miles, where it orbited a number of minutes before returning to the airfield to follow an Air Force aircraft outbound from Kirtland.

*Introduction*

This case, discussed in the Condon Report on p. 141, is an example of a UFO report which had lain in Bluebook files for years, unknown to anyone outside Air Force circles.

Immediately upon reading it, I became quite curious about it; more candidly, I became quite suspicious about it. For, as you will note on reading it for yourself, it purports to explain an incident in terms of an hypothesis with some glaringly improbable assumptions, and makes a key assertion that is hard to regard as factual. The Condon Report says:

> Observers in the CAA (now FAA) control tower saw an unidentified dark object with a white light underneath, about the "shape of an automobile on end," that crossed the field at about 1500 ft. and circled as if to come in for a landing on the E-W runway. This unidentified object appeared to reverse direction at low altitude, while out of sight of the observers behind some buildings, and climbed suddenly to about 200–300 ft., heading away from the field on a 120° course. Then it went into a steep climb and disappeared into the overcast.
>
> The Air Force view is that this UFO was a small, powerful private aircraft, flying without flight plan, that became confused and attempted a landing at the wrong airport. The pilot apparently realized his error when he saw a brightly-lit restricted area, which was at the point where the object reversed direction. . . .

The Report next remarks very briefly that the radar blip from this object was described by the operator as a "perfectly normal aircraft return," that the radar track "showed no characteristics that would have been beyond the capabilities of the more powerful private aircraft available at the time," and the Condon Report concludes, without further discussion, that: "There seems to be no reason to doubt the accuracy of this analysis."

*Some Suspect Features of the Condon Report's Explanation*

It seemed to me that there were several reasons "to doubt the accuracy of this analysis." First, let me point out that the first line or two of the account in the Condon Report contains information that the incident took place with "light rain over the airfield," late in the evening (2245–2305 MST), which I found to be correct, on checking meteorological records. Thus the reader is asked to accept the picture of a pilot coming into an unfamiliar airfield at night and under rain conditions, and doing a 180° turn at so low an altitude that it could subsequently *climb suddenly* to about 200–300 feet; and we are asked to accept the picture of this highly hazardous low-altitude nighttime turn being executed so sharply that it occurred "while out of sight of the observers behind some buildings." Now these are not casual bystanders doing the observing, but CAA controllers in a tower designed and located to afford full view of all aircraft operations occurring in or near its airfield. Hence my reaction to all of this was doubt. Pilots who execute strange and dangerous maneuvers of the type implied in this explanation do not live too long. And CAA towers are not located in such a manner that "buildings" obscure so large a block of airfield airspace as to permit aircraft to do 180° turns while hidden from tower view behind them.

*Search for the Principal Witnesses*

These facts put such strong *a priori* doubt upon the "private aircraft" explanation advanced in the Condon Report that I began an independent check on this case, just as I have checked several dozen others since publication of the Report. Here, as in all other cases, the names of witnesses are omitted; so I began my investigation through the FAA branch in Oklahoma and in California. Concurrently, I initiated inquiries concerning the existence of any structures at Kirtland back in 1957 that could have hidden an aircraft from tower view in the manner suggested. What I ultimately learned is only one of a host of examples that back up the statement I have made to many professional groups: The National Academy of Sciences is going to be in a most awkward position when the full picture of the inadequacies of the Condon Report is recognized; for I believe it will become all too obvious that the Academy placed its weighty stamp on this dismal report without even a semblance of rigorous checking of its contents.

The two tower controllers, R. M. Kaser and E. G. Brink, with whom I have had a total of five telephone interviews in the course of clarifying the case, explained to me that the object was so unlike an aircraft and exhibited performance characteristics so unlike those of any aircraft flying then or now that the "private aircraft" explanation was quite amusing to them. Neither had heard of the Air Force explanation, neither had heard of the Condon Project concurrence therein, and, most disturbing of all, neither had ever heard of the Condon Project: *No one on the Condon Project ever contacted these two men!* A half-million-dollar project, a report filled with expensive trivia and matters shedding essentially no light on the heart of the UFO puzzle, and no project investigator even bothers to hunt down the two key witnesses in this case, so casually closed by easy acceptance of the Bluebook "aircraft" explanation.

Failure to contact these two men is all the more difficult to understand when we consider that CAA tower operators who witnessed a UFO incident while actually on duty would seem to be just the type of witnesses one should most earnestly seek out in attempts to clarify the UFO puzzle. In various sections of the Condon Report, witness shortcomings (lack of experience, lack of familiarity with observing things in the sky, basic lack of credibility, and so forth) are lamented; yet here, where the backgrounds of the witnesses and the observing circumstances increase the chances of getting reliable testimony, the Colorado group did not bother to locate the witnesses. (This is not an isolated example. Even in cases which were conceded to be "unexplained," such as the June 23, 1955, Mohawk Airlines multiple-witness sighting near Utica, N.Y. [p. 143 in Condon Report], or the Jackson, Alabama, November 14, 1956, airline case, both conceded to be "unexplained," I found on interviewing key witnesses as part of my check on the Condon Report that no one from Colorado had ever talked to the witnesses. In still other important instances, only a fraction of the available witnesses were queried in preparing the Condon Report. Suggestions that the report was based on intensive investigatory work simply are not correct.)

*Information Gained from Interviews with Witnesses*

When I contacted Kaser and Brink, they told me I was the first person to query them on the case since their interrogation by an Air Force captain from Colorado Springs, who had come to interview them at Kirtland just after the incident. Subsequently, I secured the Bluebook case file on this sighting and ascertained that a Capt. Patrick O. Shere, from Ent AFB, did the interrogation on Nov. 8, 1957, just four days after the sighting.

The accounts I secured in 1969 from Kaser and Brink matched impressively the information I found in Shere's 1957 report in the Bluebook file. There were a few recollective discrepancies of distance or time estimates, but the agreements were far more significant than the small number of mismatches.

In contrast to the somewhat vague impressions I gained (and other readers would surely also gain) from reading the Condon Report version, here is what is in the Bluebook case-file and what they told me directly.

The object came down in a rather steep dive at the east end of Runway 26, left the flight line, crossed runways, taxiways, and unpaved areas at about a 30-degree angle, and proceeded southwestward toward the CAA tower at an altitude they estimated at a few tens of feet above ground. Quickly getting 7-power binoculars on it, they established that it had no wings, tail, or fuselage, was elongated in the vertical direction, and exhibited a somewhat egg-shaped form (Kaser). It appeared to be perhaps 15–20 feet in vertical dimension, about the size of an automobile on end, and had a single white light in its base. Both men were emphatic in stressing to me that *it in no way resembled an aircraft.*

It came toward them until it reached a B-58 service pad near the northeast corner of Area D (Drumhead Area, a restricted area lying south of the E-W runway at Kirtland). That spot lay about 3,000 feet ENE of the tower, near an old machine-gun calibration bunker still present at Kirtland AFB. There it proceeded to stop completely, hover just above ground *in full view* for a time that Kaser estimated at about 20 seconds, that Brink suggested to me was more like a minute, and that the contemporary Air Force interrogation implied as being rather more than a minute. Next they said it started moving again, still at very low altitude, still at modest speed, until it again reached the eastern boundary of the field. At that point, the object climbed at an extremely rapid rate (which Kaiser said was far faster than that of such modern jets as the T-38).

The Bluebook report expresses the witnesses' estimate of the climb rate as 45,000 feet per minute, which is almost certainly a too-literal conversion from Mach 1. My phone interview notes include a quote of Brink's statement to me: "There was no doubt in my mind that no aircraft I knew of then, or ever operating since then, would compare with it." Both men were emphatic in stating to me that at no time was this object hidden by any buildings. I confirmed through the Albuquerque FAA office that Area D has never had anything but chain-link fence around it, and that no buildings ever existed either inside or outside Area D in that sector. The bunker is only about 15–20 feet high, judging from my own recent observations and photos of it from the air. The Bluebook interrogation report contains no statements hinting that the object was ever hidden from view by any structures (although the Bluebook file contains the usual number of internally inconsistent and confusingly presented details).

I asked both men whether they alerted anyone else while the foregoing events were taking place. They both indicated that the object was of such unprecedented nature that it wasn't until it shot up into the overcast that they got on the phone to get the CAA Radar Approach Control (RAPCON) unit to look for a fast target to the east. Kaser recalled that a CPN-18 surveillance radar was in use at that RAPCON unit at that time, a point confirmed to me in subsequent correspondence with the present chief of the Albuquerque Airport Traffic Control Tower, Robert L. Behrens, who also provided other helpful information. Unfortunately, no one who was in the Albuquerque/Kirtland RAPCON unit in 1957 is now available, and the person who Kaser thought was actually on the CPN-18 that night is now deceased. Thus I have only Kaser's and Brink's recollections of the radar-plotting of the unknown, plus the less than precise information in the November 6, 1957, TWX to Bluebook. Captain Shere did not, evidently, take the trouble to secure any information from radar personnel.

As seen on the RAPCON CPN-18, the unknown target was still moving in an easterly direction when the alert call came from the tower. It then turned southward and, as Kaser recalled, moved south at very high speed, though nothing is said about speed in the Kirtland TWX of November 6. It proceeded a number of miles south toward the vicinity of the Albuquerque Low Frequency Range Station, orbited there for a number of minutes, came back north to near Kirtland, took up a trail position about a half-mile behind an Air Force C-46 just then leaving Kirtland, and moved offscope with the C-46. The November 8, 1957, report from Commander, 34th Air Division to ADC and to the Air Technical Intelligence Command closed with the rather reasonable comment: "Sighting and descriptions conform to no known criteria for identification of UFO's." The followup report of November 13, 1957, prepared by Air Intelligence personnel from Ent AFB, contains a number of relevant comments on the experience of the two witnesses (twenty-three years of tower control work between them as of that date), and on their intelligence, closing with the remark: "In the opinion of the interviewer, both sources (witnesses) are considered completely competent and reliable."

*Critique of the Evaluation in the Condon Report*

The Kirtland AFB case is a rather good (though not isolated) instance of the general point I feel obliged to make on the basis of my continuing check of the Condon Report: In it we have not been given anything superior to the generally casual and often incompetent level of case analysis that marked Bluebook's handling of the UFO problem in past years.

In the Bluebook files, this case is carried as "possible aircraft." Study of the twenty-one-page case file reveals that this is based solely on passing comment made by Captain Shere in closing his summary letter of November 8:

> The opinion of the preparing officer is that this object may possibly have been an unidentified aircraft, possibly confused by the runways at Kirtland AFB. The reasons for this opinion are: (a) The observers are considered competent and reliable sources, and in the opinion of this interviewer actually saw an object they could not identify. (b) The object was tracked on a radar scope by a competent operator. (c) The object does not meet identification criteria for any other phenomena.

The stunning *non sequitur* of that final conclusion might serve as an epitome of twenty-two years of Air Force response to unexplainable objects in our airspace. But when one then turns to the Condon Report's analysis and evaluation, a report that was identified to the public and the scientific community as the definitive study of UFO's, no visible improvement is found. Ignoring almost everything of interest in the case file except that a lighted airborne object came down near Kirtland airfield and left, the Condon Report covers this whole intriguing case in two short paragraphs, cites the Air Force view, embellishes it a bit by speaking of the lost aircraft as "powerful" (presumably to account for its observed Mach-1 climb-out) and suggesting that it was "flying without flight plan" (this explains why it was wandering across runways and taxiways at night, in a rain, at an altitude of a few tens of feet), and the report then closes the case with a terse conclusion: "There seems to be no reason to doubt the accuracy of this analysis."

Two telephone calls to the two principal witnesses would have confronted the Colorado investigators with emphatic testimony supporting the contents (though not the conclusions) of the Bluebook file, and rendering the suggested "powerful private aircraft" explanation untenable. By not contacting the witnesses and by overlooking most of the salient features of the reported observations, this UFO report has been left safely in the "explained" category where Bluebook put it. One has here a sample of the low scientific level of investigative and evaluative work that will be so apparent to any who take the trouble to study carefully and thoroughly the Condon Report on UFO's. AAAS members are urged to study it carefully for themselves and to decide whether it would be scientifically advisable to accept it as the final word on the twenty-two-year-long puzzle of the UFO problem. I submit that is inadvisable.

# 6

## UFO's—The Modern Myth

DONALD H. MENZEL

Myths come in a wide variety of sizes, shapes, and colors. Myths are stories, whose origins are usually forgotten, devised to explain some belief, observation, or natural phenomenon. Especially the last!

Let me remind you of a few ancient myths: Echo is a mischievous nymph who pined away for love of Narcissus until nothing was left but her voice. Earthquakes occur when a giant, chained underground beneath a mountain, tries to free himself by shaking his bonds. Lightning is a thunderbolt hurled by Zeus or Jupiter. And so on! The rain, the winds, ocean storms—all controlled by or at the mercy of some personalized deity. Man has traditionally tended to construct a myth to explain anything he cannot understand. And this is precisely the way that flying saucers or UFO's came into existence.

We usually set the date as 1947, but the phenomena responsible for the reports—or for many of them—can be traced back in history even to early biblical times. Each civilization has interpreted these natural phenomena in terms of its own culture.

The nature myths of the ancient Greeks gave way to beliefs in demons, evil spirits, the devil incarnate, witches, wizards, ogres, ghouls, harpies, fairies, fire drakes, werewolves, goblins, specters, wills-o'-the-wisp, ghosts, banshees, nymphs, elves, mermaids, leprechauns, minotaurs, centaurs, satyrs, cyclops, unicorns, and chimeras, to mention just a few. The belief in the existence of such creatures was by no means evanescent. History is full of serious claims that human beings have seen or encountered such things.

The phenomena reported as mysterious apparitions over the centuries have much in common with the modern UFO reports. People have seen some queer, luminous formation in the daytime or nighttime sky. They are frightened because they do not know what causes the apparition. They attempt, therefore, to interpret it in terms of whatever ideas may be in vogue at the time. And, almost invariably, the reporter of the incident attests to the well-known reliability and veracity of the witness. This technique is fundamental to the UFO reports.

If the literature on UFO sightings appears to be voluminous, you should examine the record for witchcraft or for devil lore. Again and again we encounter the sworn testimony of certified reliable witnesses, declaring that they recognized a particular individual as the Devil in disguise, because he failed to conceal his cloven hoof or because his forked tail protruded from his pants. And many individuals, including "innocent" children, have provided evidence that sent dozens of witches to an ignominious death.

Stories of sea serpents, dragons, and other monsters are part of our heritage, but that does not necessarily make them true. Remember the legend of the "Flying Dutchman," the ghost-ship whose master was condemned to sail the seas forever. Sighting of this fabled craft was supposed to presage bad luck. And yet such apparitions probably had a natural explanation in terms of optical mirage, as do many UFO's today.

When modern flying saucers came into being in 1947, there was a ready-made tradition waiting for them. The writer Charles Fort,[1] who died in 1932, devoted his life to collecting oddities in the news. He clipped, from magazines, newspapers, and other sources, hundreds of notes concerning seeming paradoxes of nature such as falls of stone, rains of fish or frogs, showers of ashes, inexplicable noises, and lights or flashes in the sky. Fort poked fun at scientists for lack of interest in such phenomena. He suggested that these events were the result of beings from some higher, extraterrestrial civilization "fishing for us."

The idea of flying saucers "caught on" and spread until it assumed worldwide proportions. The time was ripe for the concept, as it was for witchcraft in 1692. Man was already contemplating space exploration. So why not space travel in reverse? The view, seized upon by dozens of writers—some not too scrupulous about the facts—inflamed the credulous public with stories that were largely baseless, about flying saucers from outer space.

A new cult came into existence, with adherents who formed dozens of flying saucer clubs around the world, each with its own publication and corps of investigators: organizations such as NICAP, National Investigations Committee on Aerial Phenomena; APRO, Aereal Phenomena Research Organization; BUFORA, British UFO Research Association; GEPA (French), Groupement d'Étude de Phénomènes Aériens, and dozens of others. Each new case that hit the newspapers stimulated these believers to try to get the facts and report them. These individuals, often willfully exploited for the benefit of a few at the top, were generally sincere, honest, and hard-working. But, usually lacking proper scientific background for the studies, they made many mistakes and confused the issues.

In the early days of flying saucers, the Air Force investigations were also pretty amateurish. Their officers and personnel did not have proper background for the analysis. Frightened and confused by what they saw, or thought they saw, the military clamped heavy security on the subject, but they continued to leave control in the hands of incompetent individuals. As a Commander in the Naval Reserve, I was myself close enough to the military in those days to hear frequent behind-the-scenes rumors about the sensational findings of Projects Sign, Grudge, and Bluebook—code names for the early studies of the phenomenon.

By 1949, the evidence and conclusions were about as follows: The Air Force had collected several thousand reports of queer things in the sky. Many had come from military and commercial pilots who, presumably, were reliable and at least not likely to foster hoaxes. Objects were reported to move at speeds enormously greater than those of any known terrestrial aircraft. Similarly, the observed accelerations of these objects were far in excess of those of ordinary aircraft. They exhibited an ability to maneuver in such a way as to avoid being intercepted, so the investigators felt themselves forced to conclude that the objects were "under intelligent control." No terrestrial craft could behave in such a manner, *ergo* they must be extraterrestrial!

That, then, was the position of the military in 1949! And today, more than twenty years later, at least two scientists at this symposium are saying essentially the same thing: in brief, that they support the extraterrestrial hypothesis (ETH) simply because they cannot find any other acceptable explanation. I see nothing to justify such an assumption. In fact, I ask, is this science?

On such slender evidence James E. McDonald has flatly stated that "the problem of unidentified flying objects is, indeed, the greatest scientific problem of our times." [2] He has further urged that Congress provide, for their study, a budget that would dwarf that of NASA. J. Allen Hynek, I understand, has made a proposal to NASA for another study of UFO's. UFOlogists have warned about the danger of repeating the historic boo-boo of the French Academy about 1800 in failing to recognize that stones could fall from the sky. And yet nobody has produced a single genuine artifact—let alone "a baby UFO"—in support of the sensational, sweeping conclusions. Hynek has further admonished us to remember that there will be a science of the twenty-first, and of the thirtieth, centuries. Presumably they thus seek to refute the old-fashioned scientists who, like myself, continue to believe in the second law of thermodynamics, the impossibility of perpetual motion, the laws of conservation of matter and energy, and the laws of action and reaction.

I regard clichés as poor substitutes for scientific argument. McDonald and Hynek claim to have had considerable experience with UFO's. But their claims to authority stem from failures rather than from successes. Does mere inability to identify the stimulus of certain UFO sightings qualify one as the ultimate authority on the subject? I should think it would be quite the reverse! And why should science of the twenty-first century be specifically relevant, when the available evidence seems to establish that UFO phenomena are more closely related to seventeenth century witchcraft than to the modern world.

Briefly, my own qualifications in the field of UFO's are these: During three years of active service as Commander in the U.S. Navy during World War II, I had the responsibility of initiating and interpreting research in the field of radio propagation in general and radar phenomena in particular. I was head of the Section of Mathematical and Physical Research for Naval Communications, under the Chief of Naval Operations. I summarized the results of some of these studies in a book, originally written for the training of naval personnel in solving communication problems. There I cited startling examples of what we then called "anomalous propagation." [3]

The phenomena were indeed "anomalous" when we first encountered them. No one had foreseen that short radar waves, whose range was supposed to coincide roughly with the optical horizon, would sometimes follow the earth's curvature for thousands of miles and produce false targets that could confuse the armed forces. In the Mediterranean a cruiser shelled and reported sinking a target that later proved to be a false image of the island of Malta.[4] At last report, Malta still exists. From a distance of 600 nautical miles, a task force in the Pacific witnessed the Japanese evacuation of Kiska and ignored it because they didn't know anything about anomalous propagation. Proper interpretation of the radar record would have enabled our task force to engage and disable the Japanese fleet. I think you can see why such problems were vitally important to Naval Operations.

It was evident to many of us that a radar phenomenon analogous to optical mirage might be involved. And so I turned to meteorological optics for clues. The subject was intensely interesting. The Wave Propagation Committees of the Joint and Combined Chiefs of Staff, of which I was a member and later chairman, met weekly to discuss such problems. We directed research, both U.S. and Allied, toward a solution of this question.

The key to the problem was indeed temperature inversion: layers of cold air close to the earth's surface with the temperature increasing upward for a time. The region below the temperature maximum was called a "duct" because it tended to trap and guide the radio waves around the earth's surface. Moisture content as well as temperature proved to be important. Whether or not a radio wave remained in the duct depended on the wavelength.

I think I was one of the first scientists to obtain an analytic mathematical solution of the problem. I make no claim to originality. I just happened to recognize that the equations describing the "leaky" duct were identical with those for the escape of particles from a radioactive nucleus. The idealized solution was simple and already available in terms of Hankel functions of order ⅓.

We still encountered difficulties. Nature is never as uniform as our equations assume. The earth's surface is rough and irregular. The vertical distribution of temperature varies from point to point in an unpredictable manner. One can never have all the data necessary to achieve a full mathematical solution. So we did our best on a statistical basis. We defined a "trapping index," in terms of moisture content, temperature gradient, duct height, and wavelength.

One of the most spectacular and frightening experiences occurred off the coast of Japan, in the vicinity of Nansei Shoto.[5] Our submarines, in 1944, were carrying on extensive operations in Japanese waters. A submarine in enemy waters dared surface only at night, to replenish its air and reconnoiter the region. Our ships reported mysterious, ghostlike radar images moving at high speeds in the region. An image seemed just about to collide with the vessel when it would suddenly vanish from the radar screen. These "Galloping Ghosts of Nansei Shoto" posed an important problem for Naval Operations. It turned out that radar trapping was responsible—the result of a thin layer of cold, dry air at low levels. I wrote an article describing the phenomenon and telling how to interpret the data for the CIC (Combat Information Center) magazine.

Our researchers in the radar field turned up numerous cases of false targets and apparent trapping when the trapping index was much less than the simple theory indicated. We sometimes encountered dozens of false targets in an area where we were sure no real targets existed. Of course one can never know the temperature and moisture distribution over the entire area, but only near those regions where a sounding balloon has been sent up. From such studies we surmised that irregularities in the atmosphere—bubbles of hot air, for instance—often produce false targets even when the trapping index is less than the critical figure.

Now let us examine a few specific cases, first of radar and then of optical phenomena. I have already pointed out that, in my opinion, the views expressed by Hynek and McDonald are highly subjective. It is very hard to pin either of them down. Although both have spoken volumes on the subject, their writings are meager. For McDonald's views I turn to three sources: some of his numerous press releases, the NICAP pamphlet cited earlier, and Hearings before the Committee on Science and Astronautics, U.S. House of Representatives, Ninetieth Congress, Second Session, July 29, 1968, otherwise known as the Roush Report,[6] since it represented a symposium chaired by the Honorable J. Edward Roush of Indiana. For Hynek's views I must turn to the same Roush Report, to an earlier Congressional hearing,[7] and to two articles in popular magazines.[8]

May I digress a moment concerning the Roush Report. The contributors included mainly believers in UFO's. Five out of six of the major speakers are participating in this symposium. I was purposefully excluded from the Roush affair because, as they phrased it, they did not want to have any "extremists" participating. Roush failed of re-election, to Congress,* but he received the consolation prize of being elected to the Governing Board of NICAP, the amateur group of UFOlogists mentioned above.

If I were to discuss my own actual cases, the "believers" would accuse me of treating only IFO's: Identified Flying Objects. So I am going to confine myself mainly to cases that Hyneck and McDonald considered as unknowns, remarkable in some special way—cases that I have, in fact, studied in depth myself.

When, in the hot summer of 1952, a multitude of radar saucers invaded Washington, D.C., with concentration over the National Airport, I felt quite at home. Here were all the familiar features of anomalous propagation with its partial trapping. In confusion and fear, the authorities closed the airport and ordered aircraft from Andrews Air Force Base to try to intercept the Unknowns. The jets, directed by radar, roared into the air and found absolutely nothing. A few reported seeing distant lights but they weren't at all clear about it. The lights could have been stars, ground mirages, meteors, or false images on the retina.

The atmospheric conditions persisted for two days and were repeated five days later. Still no UFO's! No intercepts! But this failure did not discourage the UFOlogists. As one sensation monger wrote, "It was bad

* Congressman Roush was subsequently re-elected—Eds.

enough to know that UFO's were flying over Washington, but to find that they knew how to make themselves invisible was frightening!"

In the midst of this confusion I released a statement to the newspapers. I attributed the cause to a form of anomalous propagation, not fully understood perhaps but no cause for worry. No UFO's. And General Samford, a few days later, affirmed my position. Studies by the U.S. Weather Bureau and the Air Force supported my views. And so does the Condon Report. It was not surprising, I said, to expect bubbles of hot air over Washington.[9]

Now what does McDonald have to say about my views? How does he proceed? I call special attention to his methods because they are typical of his evaluations in other cases.

In the Roush Report he says: "I have interviewed five of the CAA personnel involved in this case and four of the commercial airline pilots involved, I have checked the radiosonde data against the well-known radar propagation relations, and I have studied the CAA report subsequently published on the event." He then states: "The refractive index gradient, even after making allowance for instrument lag, was far too low for 'ducting' or 'trapping' to occur." He continues in this vein for a couple of paragraphs, quoting this or that witness or authority in support of his final conclusion: "I am afraid it is difficult to accept the official explanations for the famous Washington National Airport sightings."

This kind of argument, I submit, is hardly science. The basic data, consisting of the observers' reports, obtained under conditions of panic, are clearly questionable. Those who made the reports were highly biased because they wanted to justify their original conclusions. The only hard data bearing on the question consist of radiosonde measures from several isolated points. What does McDonald know about the general propagation conditions over the entire Washington area? Nothing at all! He clearly just wants to believe that the UFO's are real and arbitrarily ignores the hard evidence. True, severe trapping did not occur. But this was one of the marginal cases of partial trapping. Harder to recognize, but the evidence is unmistakable.

Here let me review some little-known early history of the UFO's. Immediately after the end of World War II, sightings of mysterious flying objects began to multiply. More than 1,000 such reports came from Sweden alone during 1946. I heard of these through classified channels but took no part in resolving them. Air Force Intelligence, however, fully alerted, had decided that the USSR, having taken over the German rocket program at Peenemünde, was responsible for the sightings.

So when, in June 1947, Kenneth Arnold, saw "saucers" over Mt. Rainier, the Air Force was already conditioned to the idea. Here, they reasoned, were foreign experiments or expeditions. They were conditioned for another reason as well. Even then the United States was planning the famous expedition in which the U2 plane was sent on spy missions over the USSR. Could the Russians have beaten us to the punch? Were Kenneth Arnold's saucers and the following enormous rash of saucer sightings a threat to the security of the United States? Small wonder that sightings were classified and that an aura of secrecy surrounded them in the Pentagon!

I used to hear occasional juicy bits of gossip emanating from the Pentagon. I followed various reports in the news media and discovered possible natural explanations for most of the sightings. These rumors from the Pentagon reached the attention of various skillful writers who, often lacking accurate details, wove a fictional pattern to support the view that UFO's were indeed space vehicles. I am sure that many of these writers honestly believed or came to believe in the truth of their claims.

I became actively involved with UFO's in 1952. *Life* magazine had just published a fantastically sensational story called "Have We Visitors from Outer Space?" (April 7) supported by some cases the Pentagon had released. *Look* called me to ask if I had any ideas on the subject. I did and wrote two articles[10] for them which I later expanded into a book.[11] I think it is significant that not one of the original *Life* cases stands today, though believers still defend a few of them. UFOlogists never tire. And as I shoot down each of their prize exhibits, they cry, "Here's another," and wave their fantastic claims as proof of their unyielding position.

In the summer of 1952, while on a tour of active duty with the Navy, I addressed a large and enthusiastic group of officers at the Pentagon concerning my views of UFO's. I was also invited to brief the personnel of Project Bluebook. I found them much less receptive. A few were positively antagonistic, especially those who, as I later found out, had already become convinced of the extraterrestrial hypothesis, or ETH. However, I made a few converts to my views, individuals who later came to reorganize the project completely, about 1954.

Let me give you a fairly recent sighting, as reported in the *Denver Post* in January 1968. The headline read: "30 Citizens Sight UFO."

> One of the best-verified sightings of a UFO in recent months was reported in Castle Rock, a small community 30 miles south of Denver. Deputy Sheriff Weimer said about 12 "reliable citizens" [I wonder why they put "reliable citizens" in quotation marks] reported seeing a large, bubble-shaped object flying over the town between 6:10 and 6:24 P.M.
>
> Morris Fleming, director of the Douglas County Civil Defense Agency, said about 30 persons saw the object.
>
> Howard Ellis said that "all of a sudden about a dozen lights shined on me." He said the lights were "all the color of car headlights that have mud on them."
>
> Phelps said he didn't see the bubble-shaped object, but, instead, a "big, real bright light. Not a brilliant light, but a bright one." He said that the light, which moved at different speeds, seemed to be about 600 feet high and at least 25 feet in diameter.
>
> The object "shot straight up and disappeared, shooting out a couple of balls of flame," Ellis said. He thought the egg-shaped bubble was about 50 feet long, 20 feet wide and 20 feet deep.
>
> Fleming said the Douglas County Civil Defense Agency would administer a blood test to Ellis on Wednesday to determine if any "radiation or unknown or foreign matter is in his blood stream."

A remarkable and spectacular UFO; grist for the mill of the UFOlogists! And so it would be today, except for a small notice in the same paper two days later. Under the headline, "Mother of Two Young Scientists Identifies UFO," we read:

> A slightly embarrassed Castle Rock mother came forth Thursday with an explanation for the UFO viewed and reported by some 30 persons Tuesday night.
>
> The UFO, Mrs. Dietrich explained, was built by her two sons Tom, 14, and Jack, 16.
>
> "Tom learned how to make the thing in science class at school, and he was showing us how to do it," she said.

"It actually was a clear plastic dry-cleaning bag, a small one, the kind that comes on a suit jacket," Mrs. Dietrich said.

This is only an IFO, of course, but I cite it to show the unreliability of human testimony. The reports of size—50 feet long—are ludicrously wrong. But that is the trouble with most UFO reports! The records are faulty and there is no way of correcting them. Anybody reinterviewing these witnesses would not change their testimony. My chief criticism of the Air Force and their scientific consultants is their practice of giving far too much weight to the literal statements of witnesses. And the common belief that Air Force or military pilots or policemen are infallible observers is completely unjustified. For example in case after case McDonald accepts without question the statements of pilots that the UFO they have observed is solid and/or metallic. He fails completely to distinguish between the observation itself and the conclusions of the observer.

In the following case I was the observer, and McDonald has questioned either my veracity or my conclusions. Flying in the Arctic zone near Bering Strait on March 3, 1955, I observed a bright UFO shoot in toward the aircraft from the southwestern horizon. Flashing green and red lights, it came to a skidding stop about 300 feet, as nearly as I could judge, from the aircraft. Its apparent diameter was about one-third that of the full moon. It executed evasive action, disappearing over the horizon and then returning until I suddenly recognized it as an out-of-focus image of the bright star Sirius. The sudden disappearance was due to the presence of a distant mountain that momentarily cut off light from the star.[12]

McDonald, "analyzing" this sighting, characteristically and accusingly reports: "I have discussed that sighting with a number of astronomers, and not one is aware of anything that has ever been seen by any astronomer that approximates such an instance." He then questions the observation because I did not show how the index of refraction could have produced such an effect. The same procedure—interviewing selected and unidentified witnesses. I think it significant that he chose not to interview me. But, I ask, how many astronomers have seen a bright star just on their optical horizon in the clear Arctic atmosphere from an altitude of 20,000 feet? With refraction, the object would lie about 1½ degrees below the geometrical ground horizon. McDonald makes the absurd claim that such an observation would require "a peculiarly axially-symmetric distribution of refractive index, which miraculously followed the speeding aircraft along as it moved through the atmosphere, that it seems quite hopeless to explain what Menzel has reported seeing in terms of refraction effects." On the basis of a few scattered radiosonde observations, and inadequate theoretical analysis, he had rejected propagation as an explanation of the 1952 Washington sightings. Now he implies that I need detailed refractive measurements through hundreds of miles of atmosphere tangential to the earth's surface—much of it over the USSR—before he can accept my observation as valid. Nor is his statement correct that an axially symmetrical distribution of refractive index would be necessary. He was obviously unaware of an analysis I made some years ago,[13] of the "random walk" of a light beam through an atmosphere consisting of discontinuous irregular layers. You can see, perhaps, why I distrust his views and opinions. I claim—and can prove—that many of his "classical" sightings have a similar explanation, as bright stars or planets on the optical horizon. Incidentally, the UFOlogists were quick to get his message. One of the leading proponents of ETH wrote that Dr. Menzel saw in Alaska a real UFO and wasn't capable of identifying what he saw.

Another example of McDonald's scientific method is an Air Force case both of us have studied in depth. This was a sighting from the airport in Salt Lake City, October 3, 1961.[14] Harris, a private pilot, on takeoff noticed an object shaped like a silvery pencil which proved to be not a plane. It appeared to be metallic. As Harris tried to intercept it, the UFO began to move and finally, with a sudden burst of speed, faded away into the distance. During all this time ground observers reported no motion whatever.

There are many details corroborating the identification of the UFO as a sundog phenomenon, more properly called parhelia. McDonald claims that the UFO could not have been a sundog. He reported "the skies were almost cloudless." This is as though he had said, "It couldn't have been a rainbow because it had almost stopped raining." Sundogs require only a very thin layer of cirrus to become visible. Later, without explanation for his change of mind, he stated that the skies were "completely clear."

For his second point McDonald objected that a sundog would have occurred either 22° to the left or the right of the sun and at a higher elevation. On the contrary, the lower tangential arc, theoretically and practically, lies directly beneath the sun, a pencil-shaped object, at an altitude in close agreement with Harris's estimate of elevation. Moreover, parhelia, like rainbows, are centered in the eye of the observer. You can no more intercept a sundog than you can a rainbow. It is well known that parhelia possess a metallic sheen, but that does not indicate the presence of metal in the apparition. McDonald blindly accepts the observer's conclusion that he had seen a solid, metallic object.

Let me give you one final example: one of the classic sightings by Eastern Airline pilots Chiles and Whitted near Montgomery, Alabama, in July 1948.[15] They saw what appeared to be a huge, cigar-shaped, wingless aircraft. A brilliant blue glow accompanied the object and red-orange flames shot from the rear. Hynek identified this UFO as a bright meteor, and after seeing and studying the official record, I concurred with this identification. I further noted that many exceptionally bright meteors had been observed that night by amateur astronomers all over the country, because it was the date of the delta Aquarid shower. McDonald belabors me for even implying that the meteor might have been a delta Aquarid, which actually I did not do. He accuses me of glossing "over the reported rocking of the DC-3." Nonsense! There was no mention of such "rocking" in the official report. And I note that McDonald does not mention it in the Roush report.

McDonald's sole contribution to the study of UFO's—as far as I can ascertain—has been his reinterviewing of more than five hundred UFO witnesses. These interviews, clearly biased in favor of the ETH, have contributed nothing to our knowledge. They are highly subjective and have served only to crystallize the observer's earlier interpretations of his observed sighting. This is not science. McDonald and the other believers immediately consider *every* UFO to be from outer space, and they leave it to us nonbelievers to prove them wrong. I ask, should not they bring to us a better documented case than we have heard today—if they want us to take them seriously.

I confess I am much more sympathetic to Dr. Hynek's viewpoint than I am to Dr. McDonald's. Hynek is somewhat more cautious in his claims and does not come straight out in his support of ETH. Instead he implies that some vastly important scientific phenomenon may lie behind the UFO mystery. Or that there is some big secret, which he hopes to find out. Some basic discovery like that of radioactivity.

I quite understand why he would not wish to take a position that might stand in the way of such a discovery. I wouldn't want to obstruct the advance of science either. On the other hand, I think there is one far greater danger, that of fostering what the late Dr. Irving Langmuir termed "pathological science," and he included flying saucers among his items.

At one time the infamous N rays, mitogenetic radiation, and the Allison effect were as highly debated as UFO's are today. N rays were supposed to be a mysterious radiation emitted spontaneously by various metals. After being passed through a spectroscope whose lenses and prisms were of solid aluminum, these rays impinged on the dark-adapted eye, which detected them as flashes of visible light. Nearly one hundred papers on N rays were published in *Comptes Rendus* in the first half of 1904 alone. And the French Academy awarded Blondlot the Lalande prize of 20,000 francs and its Gold Medal for the "discovery." The irrepressible R. W. Wood cleverly exposed N rays as a figment of Blondlot's imagination—self-delusion. The "flashes" were purely physiological, an optical illusion, a natural reaction of the unreliable human retina. This phenomenon is undoubtedly also responsible for many UFO reports. Apollo astronauts, blindfolded in orbit, have reported seeing such flashes. Some scientists have attributed them to the stimulating effect of cosmic rays. But the physiological explanation is more probable.

Mitogenetic radiation was supposedly electromagnetic energy emitted by the roots of growing plants. And through the Allison effect one supposedly could detect the presence of isotopes of rare substances. These, again, were the results of self-delusion, with purely subjective detection. All pathological science!

I submit that McDonald's interviews of more than five hundred people who have reported UFO's have no scientific validity whatever, except to confirm his well-known bias in favor of ETH and against the Air Force and myself and other nonbelievers. Similarly, Hynek's indexes of "credibility" and "strangeness" are equally subjective. Study of them may throw some light on Dr. Hynek but they are unlikely to contribute much to the UFO problem.

At this point I must reveal that between 1962 and 1970 the Air Force has increasingly used my services as a consultant in analyzing UFO reports. Most of the cases they have sent me during the past few years have been ones that Hynek proved unable to solve: those he listed as "unknowns." Furthermore, I have solved the majority of such cases, so that the Air Force files no longer list them as "unknowns." I shall mention only two of these cases, identifying them by date and location, but using a fictitious name for the observer, as the Air Force requested.

The first case occurred April 3, 1968, near Cochrane, Wisconsin. Betty, driving at 8:15 P.M. on the highway with her ten-year-old son, noticed a luminous orange object hovering overhead. It was shaped, she said, like a boomerang or a croissant. It was fuzzy and seemed to be covered with "angel hair." Suddenly the car engine died and the lights went out. She shut and locked the windows and it got very hot in the car. Then she started the car after the UFO disappeared. Both she and the boy were very frightened. The sky was partially cloudy. The moon was crescent.

Hynek's decision: "Unknown."

The second case was that of Yellow Springs, Ohio, August 15, 1968. Alice, driving a convertible with the top down, became aware of a bright light directly overhead. It seemed to be spinning. When she stopped the car, the UFO stopped; when she started up the UFO followed. She did this several times. Badly frightened, she drove home as rapidly as possible and called her husband and husband's parents to look at it.

Alice testified that the object was fuzzy, as if it had a fog or mist around it. It seemed to oscillate jerkily back and forth, so she concluded it couldn't be the moon. Her husband said at first it was just the moon, but changed his mind when he, too, noted the jerky motion. Others confirmed that statement. Curiously, the moon was in the sky at the time—though, since no one reported seeing it, Hynek decided that the sky must have been cloudy or partly cloudy so as to obscure the moon.

Hynek's decision: "Unknown."

I analyzed both of these cases independently and stated that, in both of them, I thought the observer had seen the moon through variable haze. The apparent failure of the electrical system had no significance, other than the possibility that Betty, frightened and hysterical, accidentally killed her engine. Her use of the phrase "angel hair" clearly denotes previous conditioning in UFO lingo. "Angel hair" to a UFOlogist signifies falling clumps of fine massed threads accompanying certain kinds of UFO activity.[16] Actually, angel hair appears to be special webs spun by the parachute spiders, in which they lay their eggs. The wind catches them up and carries them to great heights. In the sun they glint like spun silver. Here, however, I am suggesting only that fog or haze produced the illusion. Hynek considered the moon explanation but rejected it because he could not see how people could be so dumb as not to recognize the moon when they saw it. He forgot that Alice's husband first identified it as the moon and then rejected it because it moved in jerks.

These points the Air Force has overlooked from the very first. I have often heard Hynek say, "Stars and the moon don't cavort over the sky." That statement is true, but it fails to take into account a well-known physiological phenomenon called autokinesis. The apparent motion results from uncontrollable irregular movements of the eyeball. Minnaert, in his delightful book *The Nature of Light and Colour in the Open Air* (New York: Dover, 1954) mentions the swinging stars and gives a reference where "simultaneously three people saw the moon dance up and down for thirty minutes." And as for the moon's stopping or moving with the car, that also is a well-known optical illusion. When I was a child I remember watching the moon from a train and wondering how it managed to keep up with the moving cars and stop when we came to a station.

Anyway, these cases are marked as "The Moon" in Air Force records. And you may conclude that I am much more skeptical than Hynek concerning the reliability of observers.

Failure to recognize phenomena of the human eye is a major defect of the Air Force UFO questionnaire. I don't know who is at fault, but the questionnaire seems cleverly designed to avoid asking the most vital questions and to get the wrong answers. In response to my repeated criticisms, the Air Force asked me to suggest revisions for the new printing. I spent several weeks, detailing the revisions and giving my reasons therefor. But they adopted only a few of my suggestions, rejecting the remainder because they considered them to be an invasion of the privacy of the individual.

The original questionnaire determined whether or not the person was wearing glasses, but it did not find out whether a person who was not wearing glasses was supposed to wear them. I wanted to know how long it had been since the witness had had an eye examination. I even wanted to know the nature of the correction. But that was an invasion of privacy. I am willing to bet that both Alice and Betty needed glasses, though they were not wearing them.

Such occurrences are by no means new. The astronomer Simon Newcomb tells in his autobiography of an experience he had in 1860, returning across Minnesota from observation of a total eclipse of the sun. Some officers from Fort Snelling claimed they had seen a star that behaved in most surprising fashion. It rose in the east, then turned north, and finally set near the north. They showed it to Newcomb, who immediately identified it as Mars. Several hours passed and then one of the officers pointed to a bright star just on the horizon, saying, "There it is, setting just now." Newcomb identified the second star as Capella, rising, and pointed out Mars, by then inconspicuous near the meridian. Newcomb commented that "the men who saw it were not of the ordinary untrained kind, but graduates of West Point, who, if any one, ought to be free from optical deceptions." [17]

At a Congressional hearing in April 1966, Hynek said: "I have set aside for further study some 20 particularly well-reported UFO cases which, despite the character, technical competence and number of the witnesses, I have not been able to explain. I have done this to illustrate that neither I nor the Air Force hide the fact that there are unexplained reports, and to illustrate also that the Air Force does not maintain, contrary to some public opinion, that reporters of UFO's are lacking in intelligence or are objects fit only for ridicule." [18]

Intrigued by this statement and looking on it as a sort of challenge, I wrote to Hynek, asking him whether he intended to sit on this evidence, regarding it as his personal property, or whether he would be willing to make it available to me. After a long delay I received eleven of his cases from the Air Force files. I found many of them lacking in solid detail. One of them had the wrong weather data attached. Nevertheless, within the limits of the available data, I was able to suggest reasonable solutions for all eleven cases.

One in particular gave me some trouble at first. It related to a sighting on July 20, 1964, from Yachats, Oregon, of a starlike object moving northeast in a straight line. Its motion was not uniform in that it seemed to pause momentarily in its path. Except for this wavering, it behaved like a satellite, although Echo II was too close to the horizon to be visible. After many checks, it turned out that the object was indeed Echo II, but the recorded day was wrong, because of an error in conversion of Pacific time to Greenwich time and back. The wavering was explicable as the reverse of autokinesis: autostasis, irregular following by the eyeball.

A second case particularly appealed to me. On February 6, 1966, a child, going to the bathroom in the middle of the night, turned on the light and in so doing awakened his parents. The light suddenly went out and the father got out of bed to investigate. He happened to glance out of the window and was surprised to see a pulsating, reddish glow that moved irregularly over the sky and shortly faded out. Next morning it was determined that a blown transformer had caused the light to go out. But that probably had nothing to do with the UFO. The father, dark-adapted from sleep, caught the bright light full in his eye. The result was an after-image, which drew his attention as he passed the window. Simple, but the Air Force questionnaire provided no basis whatever for recognizing an after-image. In fact the entire Air Force questionnaire is based on the premise that a UFO is always a solid, material object. No wonder that so many of them have been classified as "unknowns."

When Condon asked Hynek for these cases in February 1968, Hynek refused on the basis that, by then, the Colorado investigation would shortly come to an end, and he did not want to run the risk of having potentially valuable data rendered useless or jeopardized for future work, through careless processing of the material. Hynek emphasized the desirability of studying all the cases collectively, with the hope of finding relevant patterns of similarity between them. In particular, he objected to the method of treating each case separately and individually. He stated, "It is clear that each case, taken by itself, like a lone duck in a shooting gallery, can nearly always be shot down by an ad hoc, frequently Menzelian approach."

I feel rather honored to become an adjective. But I simply cannot understand how Hynek feels that the cases can be "shot down" individually, but not collectively. Each case is a separate item. It seems highly dangerous to suppose that one can add data from another case, unless one is absolutely sure they concern the same phenomenon.

Hynek speaks of wanting to know if there is a real "signal in all the noise," by which I assume he is asking if even a few of the UFO reports relate to some entirely new phenomenon, ETH or otherwise. Suppose, for example, you had one hundred phonograph records of Caruso singing *Il Trovatore,* all so badly scratched that you cannot even recognize the music. If you were to play these hundred records synchronously on a hundred record players and make a recording of the combined output, you would increase the signal-to-noise ratio by a factor of 10, the square root of the number of records, according to information theory.

But such an analysis is valid only if you are sure that the records are of the same piece. Remember that UFO records are unlabeled. I can list more than one hundred entirely different stimuli that can produce the observation. Even certain gross similarities of two records from a given geographical area at about the same time do not guarantee that the stimuli are the same. Trying to analyze such randomly selected records by information theory is something like superposing phonograph records of several dozen different types. We know how "noisy" the records are from the incompatibility of sightings of the same event, such as the plastic balloon of Castle Rock, cited earlier, or the spectacular satellite re-entry of March 3, 1968, over Indiana and Ohio and as far south as Tennessee. One witness claimed to have seen "windows with faces behind them!" Another swore that one object landed just over the next hill. How can anybody expect to analyze reports as "noisy" as those. And of what possible value can a reinterview of these witnesses be?

What, then, are the UFO's? I repeat, there are hundreds of varieties.

But perhaps I should be a little more specific, though the number is so great that I shall have to confine myself to selected examples rather than give a complete listing.

A. Material objects

1. Upper atmosphere
   - meteors
   - satellite re-entry
   - rocket firings
   - ionosphere experiments
   - sky-hook balloons
2. Lower atmosphere
   - planes
     - reflection of sun
     - running lights
     - landing lights
   - weather balloons
     - luminous
     - nonluminous
     - clusters
   - clouds
   - contrails
   - blimps
     - advertising
     - illuminated
   - bubbles
     - sewage disposal
     - soap bubbles
   - military test craft
   - military experiments
     - magnesium flares
   - birds migrating
     - flocks
     - individual
     - luminous
3. Very low atmosphere
   - paper and other debris
   - kites
   - leaves
   - spider webs
   - insects
     - swarms
     - moths
   - luminous
     - (electrical discharge)
   - seeds
     - milkweed, etc.
   - feathers
   - parachutes
   - fireworks
4. On or near ground
   - dust devils
   - power lines
   - transformers
   - elevated street lights
   - insulators
   - reflections from windows
   - water tanks
   - lightning rods
   - TV antennas
   - weathervanes
   - automobile headlights
   - lakes and ponds
   - beacon lights
   - lighthouses
   - tumbleweeds
   - icebergs
   - domed roofs
   - radar antennas
   - radio astronomy antennas
   - insect swarms
   - fires
   - oil refineries
   - cigarettes tossed away

B. Immaterial objects
   1. Upper atmosphere
      - auroral phenomena
      - noctilucent clouds
   2. Lower atmosphere
      - reflections of searchlights
      - lightning
        - streak
        - chain
        - sheet
        - plasma phenomena
        - ball lightning
      - St. Elmo's fire
      - parhelia
        - sundogs
      - parselene
        - moondogs
      - reflections from fog and mist
        - haloes
        - pilot's halo
        - ghost of the Brocken
      - mirages
        - superior
        - inferior

C. Astronomical
   - planets
   - stars
   - artificial satellites
   - sun
   - moon
   - meteors
   - comets

D. Physiological
   - after-images
     - sun
     - moon
     - reflections from bright sources
     - electric lights
     - street lights
     - flashlights
     - matches
       (smoker lighting pipe)
   - autokinesis
     - stars unsteady
     - stars changing places
     - falling leaf effect
     - autostasis
       (irregular movement)
   - eye defects
     - astigmatisms
     - myopia (squinting)
     - failure to wear glasses
     - reflection from glasses
     - entoptic phenomena
       - retinal defects
       - vitreous humour

E. Psychological
   - hallucination

F. Combinations and special effects

G. Photographic records
   - development defects
   - internal camera reflections

H. Radar
   - anomalous refraction
   - scattering
   - ghost images
   - angels
   - birds
   - insects
   - multiple reflections

I. Hoaxes

This listing is minimal and highly abbreviated. But none of the questionnaires I have seen, either those of the Air Force or those of the many amateur groups, was designed to detect, separate, and identify the majority of the various phenomena listed above. I have urged that the Air Force questionnaire ask the question: "What natural phenomenon did your sighting most closely resemble?" then, the clincher: "Why do you feel that UFO was not this phenomenon?" The Air Force did not accept my suggestions. Where I independently had a chance to follow up the question, I found many times that the reason given was inadequate. For example: "It couldn't have been a plane because I couldn't hear the engine, or because the light was too bright." Or: "It couldn't have been a meteor because it was moving up, and meteors fall down." Here is a curious conflict of reference coordinates between the observer and the meteor. Many persons fail to realize that a meteor, actually falling, can appear to move up—that is, away from the observer's horizon.

In conclusion I want to point out that, in my opinion, the question of whether planets of our solar system or elsewhere have intelligent life on them is irrelevant. Nor am I denying the possibility that someday we may actually experience visits from outer space. My point is that the UFO reports to date do not represent extraterrestrial activity in any form. I confidently predict that no amount of investigation will bring evidence in support of the extraterrestrial hypothesis.

It is well established that reports of UFO sightings come and go in waves. Many people seem to have the idea that UFO stimuli also ebb and flow. UFOlogists seem to think the way to solve the UFO mystery is to have thousands of task forces all over the world ready to ride at the report of a UFO, with the hope of getting there before it vanishes. But this is not the way it happens at all. There are dozens of stimuli around all the time. I can't walk around the block without seeing at least one and sometimes several of the basic stimuli that people have reported from time to time as a bona fide UFO. Why don't people report them, then? The answer is simple and obvious. When UFO's are in the news people look for them and see them. As the publicity subsides the reports subside.

I was delighted at the news released by the Air Force on December 18, 1969, that they were giving up all further collection and analysis of

UFO reports. This is high time. Twenty-two years of study have yielded essentially nothing of positive value, from the viewpoint of science, and from the standpoint of military intelligence.

The scientific world should be highly grateful to Dr. Edward U. Condon of the University of Colorado, who undertook, in the public interest, an independent and unbiased study of UFO's. The Condon Report [19] is by no means exhaustive or free from errors. No finite study could have achieved perfection. If Dr. Condon did err in judgment, it was in taking the considered risk of employing some individuals known to be ardent believers of UFOlogy. This action was in keeping with his well-known record of fairness. It is not surprising that some of these persons bitterly criticized both Dr. Condon and the study, when the negative character of his conclusions became known. But Condon's book deserves our support as well as our gratitude. I heartily endorse the report and concur with its general findings.

The Condon Report has cooled off UFO interest, and reports are at their lowest ebb in years. The UFO groups, fearful of having to fold up completely, are desperately trying to generate new interest in this topic. On good authority, I understand that UFOlogists look hopefully to this symposium, sponsored by the AAAS, as a means of rekindling public interest in this field. For this reason, I at first declined the invitation to participate. I did not relish the idea that the press would likely feature the sensational claims, however absurd, of the UFOlogists. But Walter Roberts convinced me that I should change my mind, and lend at least some semblance of balance to this symposium.

I was not surprised to hear that NICAP also expressed delight at the Air Force announcement. It leaves them in undisputed possession of the field. Clearly, they will make one last effort to secure government support for their own UFO studies, designed to justify their undying faith in the extraterrestrial character of UFO's, before the subject declines into the oblivion it deserves. Do not take these amateur groups lightly. They can do considerable harm to science with their vociferous demands for costly government studies. I hope the silent majority will speak up against this situation. I think the Condon Report will hold the barrier. I am concerned to learn that this report is disappearing from libraries around the country at a rate far greater than one would expect for a book that costs only $2.00. I wonder whether this might not be an attempt at suppression by various individuals who regard it harmful to the cause of UFOlogy.

I do predict however, a continued decline of public interest in UFO's. The people seem to have taken up a new cause: Astrology. It has a similar scientific basis and fulfills a similar need in human desire. Within the two years that have elapsed since the AAAS Symposium and the printing of this book, the number of reports has dwindled almost to the vanishing point. Most of the UFO societies have quietly folded. Only a few die-hards and sensation-mongering journals still urge support for the moribund ETH.

The government should withdraw all support for UFO studies as such, though I could advocate the support of research in certain atmospheric phenomena associated with UFO reports. I further predict that scientists of the twenty-first century will look back on UFO's as the greatest nonsense of the twentieth century.

And now, as UFO's gradually slide back into mythology, I leave you with what I consider an apt quotation from Shakespeare's *Merchant of Venice:*

> All that glisters is not gold;
>
> . . . . .
>
> Gilded tombs do worms infold.

## APPENDIXES

### 1. The Papua Case, June, July, and August, 1959

One of the UFOlogists' favorite cases occurred during June, July, and early August of 1959. Many observers reported seeing one or more UFO's, sometimes simultaneously. The chief observer was William B. Gill, a priest in charge of the Anglican mission at Boianai, Papua. Of the thirty witnesses who saw the event, all but Father Gill were Papuans. Six of these were teachers; the remainder were children. Twenty-four signatures appear on a sheet apparently testifying to the reality of the phenomenon. The practice of letting the Papuans choose their own names makes some of the signatures appear slightly odd. Thus we find Ananias, Kipling, Love Daisy, and Annie Laurie.

Most of the sightings took place in the early evening, after sunset, though some of the UFO's were seen in the early morning hours. The first sighting, which occurred on June 26 and to the report of which most of the witnesses signed their names, is not described as clearly or as graphically as the one on the following night. Since the two sightings were evidently very similar, I give the second in Father Gill's own words.

A large UFO was first sighted by Annie Laurie at 6:00 P.M. in apparently the same position as the one last night, only that it seemed a little smaller, when I saw it at 6:02 P.M. I called Ananias and several others and we stood in the open to watch. Although the sun had set, it was quite light for the following 15 minutes.

We watched figures appear on the top—four of them. I had no doubt that they were human. It was possibly the same object that I took to be the "Mother Ship" last night. Two smaller UFO's were seen at the same time, stationary. One above the hills, to the West, another overhead.

On the large one, two of the figures seemed to be doing something near the center of the deck. They were occasionally bending over and raising their arms as though adjusting or "setting up" something that was not visible. One figure seemed to be standing, looking down at us (a group of about a dozen).

I stretched my arm above my head and waved. To our surprise the figure did the same. Ananias waved both arms over his head, and then two others apparently lost interest in us for they disappeared below deck.

"At 6:25 P.M. two figures re-appeared, to carry on with whatever they were doing before the interruption (?). The blue spotlight came on for a few seconds twice in succession.

The two other UFO's remained stationary and high up, higher than last night (?) and smaller than last night.

At 6:30 P.M. I went to dinner.

At 7:00 P.M. the number one UFO was still present but appeared somewhat smaller. The observers went to Church for Evensong. 7:45 P.M., Evensong over and sky covered. The visibility was very low.

At 10:40 P.M. a terrific explosion occurred just outside the Bishop's House. Nothing was seen. It could have been an electrical atmospheric explosion as the whole sky was overcast. At 11:05 P.M. a few drops of rain fell. This may or may not have anything to do with the UFO. The explosion seemed to be just outside of the window, *not* an ordinary thunder clap, but a penetrating "earsplitting" explosion. It woke up people in the Station.

Several days later Father Gill wrote the following addendum:

Have been having further experiences lately with the UFO. On Saturday night I counted one large and seven small UFO's; on Sunday one large and two small; on Monday one large and four small.

Believe it or not, Ananias, Mission boys and I exchanged hand signals (before dark Saturday) with the occupants of the "Mother Ship" a little after 6:00 P.M. There is no doubt that she is occupied by at least four men. Assuming that the men are the same size as the average of those on earth, I have worked out the size of the Mother Ship. The top deck is about 20 feet in diameter. The bottom deck is roughly 35 feet. The distance of the operation, therefore, as we have sighted, have been at the highest altitude 2500 feet; the lowest altitude 450 feet. . . . No human activities have been observed on the smaller UFO's. They seem to be disks rather than saucers.

Various other people reported the sighting, describing it as a "bright light," which often changed color from green to red. The object remained in the western sky, slowly descending until it finally disappeared behind a bank of clouds. As I said earlier, the vivid character of this sighting has made it the darling of the UFOlogists. There is no question, of course, of the integrity of Father Gill. But how reliable he is as an observer is quite another question. Although many of his associates and students signed the paper that he wrote, we have no assurance that they really knew what they were signing. All of them, undoubtedly, saw something bright in the sky. And if that bright thing, whatever it was, so impressed their priest, they would certainly have signed and attested to anything. One can hardly term them independent, unbiased observers.

Certainly Father Gill was familiar with the UFO phenomenon. He certainly used the UFOlogist jargon, such as "Mother Ship," and indicated that he had become "convinced" of their reality. Perhaps, if I had seen the same phenomenon, I too would have become convinced. On the other hand, I think there is a reasonable possibility that I might have found the solution immediately.

One thing that, to me, seemed a trifle peculiar was the laconic statement: "At 6:30 P.M. I went to dinner." Here was a man reportedly seeing one of the most spectacular phenomena of history, at least in terms of his own explanation. His native curiosity must have been fairly low, if he allowed the pangs of hunger to pull him away from the view of a "Mother Ship" peopled with beings from outer space who were waving and apparently trying to communicate with the observers on earth. So he went to dinner and when he came out everything had clouded over.

What, indeed, could Father Gill have seen? There is one thing lacking from his reports that provides a clue. The planet Venus was then an evening star, near its maximum brilliance. This planet has frequently been reported as a UFO. The earth's atmosphere frequently causes Venus to twinkle and change color, running the whole gamut of the spectrum from red to green. This planet must have been roughly in the position indicated by Father Gill. And yet he never even mentions it as a point of reference. The question occurred to me quite early in my study of this fantastic case: Could Father Gill have been viewing the planet Venus?

There are some obvious objections to this simple solution, which have also occurred to the UFOlogists. Planets don't appear to have men standing on them. Planets do not send out search lights. So, how to reconcile these observations with the planet Venus?

Father Gill's drawing shows little more than a bright oval, with four legs or other supports at the bottom and from one to four men up above. He also reported the beam of the searchlight, shooting upward from the UFO, but this could easily have been the effect of clouds. And, although several children signed their names, we have no assurance that the drawing as executed by Father Gill in any way resembles what they themselves saw or thought they saw. Here was the Reverend, tremendously excited about something. He reported that the Mission boys made audible gasps (of either joy or surprise, perhaps both). Is it not likely the boys were surprised to see their leader so excited about a phenomenon that might not have been in the least mysterious to them?

A number of the reports made throughout July and early August actually mention the planet Venus. However, the fact is that Venus and the UFO were never reported visible simultaneously during July and early August when Venus even increased slightly in brightness. The other objects reported could be identified with other planets or first-magnitude stars. For example, Jupiter in the morning sky was quite brilliant, though not as bright as Venus.

One evening, several years ago, while watching Venus in the western sky, I suddenly thought of a possible explanation of Father Gill's sighting. Suppose, I thought, that the priest, perhaps unknown to himself, has considerable myopia and astigmatism in his eye. I am slightly myopic myself and remember how the stars appeared to me before I started to wear glasses at the age of twelve. The bright stars always looked big and the fainter stars smaller. But they were always fuzzy blurs.

I decided that I could simulate both the myopia and astigmatism with the aid of lenses. From a large selection of spectacle lenses, which I occasionally use in my astronomical experiments, I chose a positive lens that had a certain amount of astigmatism. I removed my spectacles and inserted this lens in one eye, something like a monocle. To my delight, but not to my surprise, Venus and the other stars flattened out and became saucers. A person who has myopia can improve his vision somewhat by squinting. The effect is something like that of the iris diaphragm of a camera. The smaller the aperture, the sharper will be the picture, especially if the image is slightly out of focus.

But there is still another phenomenon involved, which any astronomer will easily recognize. The out-of-focus image of the star on the retina is really an image of the lens of the eye. As I squinted at Venus, the planet pulsated, tended to change shape, an effect caused by the eyelids, which narrowed the aperture of the pupil. But the remarkable thing was an appearance caused by the slightly out-of-focus images of my eyelashes. With a little imagination, these luminous projections, extending above and below the saucer-shaped image, appeared like the men or legs of Father Gill's drawing. The slight movement of my eye, up or down, caused the "men" to move around. The slight irregularities on the "hairs" of the lashes, perhaps dust or moisture, could easily be interpreted as activity of the "beings" inhabiting the saucer. In brief, by a simple experiment, I had reproduced most of the phenomena reported by Father Gill.

I can see that this explanation is mere speculation on my part. Moreover, I am disregarding the signatures of the witnesses who reportedly confirmed what Father Gill had seen. I have made a number of attempts to contact Father Gill, who, I am sure, is an honest man who would like to contribute to the resolution of this strange apparition, unique in all of the history of flying saucers. None of my letters was returned, but I have never received any reply. Hence my solution must remain conjecture. However, I'll argue that my explanation is a reasonable one and much more probable than the alternative, that UFO's were flying over Papua, trying to communicate with the inhabitants.

There were miscellaneous additional reports by other observers, rarely more than two, of UFO's dashing across the sky. The observers themselves admitted that what they saw could have been ordinary meteors. During the height of the phenomena there was a mysterious daytime sighting on July 21, according to the Reverend Norman E. G. Cruttwell, associated with another of the Anglican missions. Let me quote from his report.

> It appeared over the hill to the West of the Station and traveled Eastward at an angle of about 30 degrees above the horizon. It traveled quickly, faster than an airplane, but without a sound. It was visible far less than a minute.
>
> When it first appeared it was a point of white light in the sky, like a star. The sky was bright and almost cloudless. As it approached, it appeared to increase rapidly in size and to take on an elongated shape. And at its nearest point it had the appearance of a disk, shining silver in the sun and somewhat smaller in appearance than the sun's disk. It appeared to have a darker rim, giving a ring-like effect. It then receded to the East and finally faded out into the distance as a point of light. It apparently wavered slightly in its course as it receded as if effected by the wind.
>
> It was seen by six (Papuans), two teachers and many children. They sent down for the Reverend N. E. G. Cruttwell, who was indoors, but they wasted too much time, and by the time he arrived in a clear place the object was out of sight, though the children and teachers were still standing staring at the sky. It did not return.

The object was also seen at approximately the same time at Koyabagira and Giwa, villages 15 and 12 miles away respectively.

If one discounts the uncorroborated reports of simultaneous sightings, there are many explanations of the phenomenon. A bird flying high, some sort of a seed pod, or almost any kind of wind-borne debris could have produced an effect of this sort. And clearly the witnesses have been conditioned, by virtue of the publicity of the Father Gill sightings, to report almost anything.

All of the evidence points directly toward planetary and stellar objects as the source. The reported twinkling, the changes of color, and even the general motion which was always toward the west, concur with this identification. I have previously shown that oscillations of the human eye can cause the stars and planets to "cavort" around the sky. And I think that this phenomenon, known technically as autokinesis, was responsible for some of the reports. Light refraction, the changing colors caused by optical interference of light waves, and reflections from clouds account for most of the other reported effects.

Few people realize how imperfect their own eyes can be. To prove my point, I occasionally ask someone to look at a bright star or planet and tell me how many "points" he can see. I then ask him to draw a picture of the star, with the various extensions. He often will not believe me when I tell him that, if his eyes were perfect, the star would look like a single point, without any projections at all. But I can usually convince him by telling him either to lie on his side or at least to turn his head to one side or the other and see that the "points" or projections rotate with his head. Everybody has some optical defects of this type, which can account for the reports of "searchlight beams" and other phenomena. Even a transient mote, such as a dust speck, on the retina can produce image distortion.

I do not claim that I have completely solved Father Gill's Papuan sighting, but I have provided a reasonable explanation in terms of well-known phenomena. It is significant, and hitherto unrecognized, that, as Venus drew rapidly toward the sun and decreased sharply in brilliance, the reports of UFO's from Papua similarly decreased. I think, moreover, that this particular sighting represents an interesting psychological phenomenon. It shows the effect of one observer upon another. When someone in authority claims to have seen something, those who think they should be observing the same phenomenon express blind agreement. A follow-up of the Papuan case, with interviews of the witnesses who signed, would indeed be interesting. If that were done by anyone other than a UFOlogist, I think we would find that the individuals did not see the little men. If, perchance, they did, I would like to see a report of an oculist concerning their eyes and vision.

## 2. Clearwater, Florida, July 4, 1964

This case is one of the best reported and most widely observed in all of UFO history. The complete Air Force record, of which I have a copy, contains almost five hundred pages. This case was never reported or discussed in the flying saucer literature because quick action on the part of the Air Force completely solved the mystery. The case is significant chiefly because of the large number of observers.

In the early evening of July 4, 1964, an enormous fireworks display was in progress. As a result, hundreds of people were looking at the sky. On July 6, the *Clearwater Sun* reported:

> Six mysterious red lights which appeared over the Gulf at Clearwater Beach Saturday night remained a mystery today, but persons who observed the lights were asked to write descriptions of what they saw for the Air Force.
>
> Capt. R. H. Henry, Public Information Officer at MacDill Air Force Base, Tampa, was in Clearwater to view the fireworks Saturday night. He saw the lights but didn't attach any significance to them.
>
> He asked that people who viewed such phenomena send descriptions to the Air Force in care of his office.

Altogether, the Air Force received sixty-six reports. As in all such cases, they differed significantly from one to another. Some said that the lights were blinking and flickering. Others that they remained constant in brightness. There was enormous fluctuation in the estimates of the altitudes of the lights, ranging from less than 1,000 feet to more than 20,000 feet. The spacing between the lights ranged from 10 feet to as much as 200 feet. The number of lights reported ranged from as few as three to as many as seven. The colors were variously observed as red, white, green, blue, or combinations of these. The one factor that most people agreed on was that the lights seemed to lie pretty much in a straight line. Several observers commented, perhaps facetiously, that they'd heard someone say the Martians were arriving.

The following letter, addressed to the Director of Intelligence, 12th Technical Fighter Wing, MacDill Air Force Base, Tampa, provided a complete solution.

> On July 4 about 8:00 P.M., I took off from Clearwater Air Park in my Cessna 170-A, with five small plastic parachutes, with a 10-minute railroad flare attached to each, spikes removed. I took off for the North West and climbed to 500 feet, then made a 90° turn to the left out of the flight pattern. Still climbing, I headed to the South-West over Clearwater to about Clearwater Beach, then turned North to about Dunedin or Ozona, then South again over the Islands to Belleair Beach, again turning North to about 5 miles West of Clearwater Beach, at which time I released the five flares about 10–15 seconds apart at the altitude of 10,000 feet, after which I circled the flares until they went out at about 7,500 feet. One flare dropped after it had fallen about 1,000 feet. After the flares went out I returned to Clearwater Air Park and landed.
>
> There were no planes visible at the time of the flight and only one boat light or beacon visible about 20 miles West. It was a very clear night.
>
> The reason for the flares was to add to the 4th of July festivities with something different, but [I] had not intended to stir up such a commotion in the newspapers and did not think the Air Force would be so concerned. I am truly sorry if I have caused any inconvenience to anyone.
>
> The time of the climb was about 25 minutes. The time of the drop was approximately 8:30 P.M.
>
> James G. Mercer

The *Clearwater Sun* carried the story on July 9.

Of some significance is the fact that less than one-third of those reporting gave the number of lights as five. Only 5 percent recognized them as flares. About 5 percent indicated that a parachute was involved. None of the reported separations was anywhere nearly correct. The minimum separation possible was about 1000 feet, an estimate made from the known speed of the plane and the interval between successive drops. About 30 percent of the observers recognized that a plane accompanied the lights.

This sighting shows the tremendous interest that UFO's can generate, even when the sighting is so obviously connected with the fireworks display in progress.

### 3. The Spectacular UFO's of March 3, 1968

Another sighting the UFOlogists would like to forget was one of the most spectacular on record. The event occurred at about 9:50 P.M., Central standard time. The UFO's were reported from at least nine states: Indiana, Kentucky, Massachusetts, New York, Ohio, Pennsylvania, Tennessee, Virginia, and West Virginia. Hundreds of people observed one or more brilliant fiery objects streaking across the sky, sending out showers of sparks, and leaving bright trails behind them. Hundreds of people made detailed reports of their sightings. Air Force records of this event, of which I have a complete copy, run to more than 400 pages.

By far the most detailed and graphic report came from a woman in Tennessee, whom I shall call Marie, because that is not her name. She, her husband John, and the mayor of the town were the observers. Following is her complete report, edited only to remove identifying names.

Sunday, March 3, 1968. A number of us had enjoyed dinner at the mayor's home in———. About 8:43 P.M., C.S.T., the mayor, John, and I left the house and walked through the parking lot where the three of us stood talking. I saw a light traveling in the sky a little above the southwest horizon. This light seemed only a bit larger and brighter than a star and it seemed about the same color as a star.

As I yelled to the mayor and John to look, the light became brighter and larger. While I was observing this "traveling light" from a great distance, it did not look to me that it was traveling in a flat trajectory. Rather, it seemed to travel in a slight arc and, at this point of flight, I began to note the "orangish-colored" trail of light behind the "starcolored" light. John asked, "You do know what we're seeing, don't you, Mayor?" John's talking was an annoying distraction to me while I was trying to listen for some sound, so I bossed loudly, "Hush your mouth!"

The three of us stood silent, almost motionless, and very much in awe as we realized that the "thing" was headed our way and was coming surprisingly near us! There were some leafless trees in the yard that partially obstructed our view for a moment. Then—IMPACT!!!—The "impact" I am referring to is the impact on my emotions, for with breathtaking suddenness, the "thing" was nearly overhead and seemed to be quite large and close! To be more explicit, the "thing" looked like it was headed directly over the far corner of the mayor's house!

It was shaped like a fat cigar, in my estimation. I was impressed that it seemed of considerable size, the size of one of our largest airplane fuselages, or larger. (The mayor thought it was smaller than my estimation.)

It appeared to have square-shaped windows along the side that was facing us. I remember the urge to count the windows, but other details flashed in view and my curiosity made me jump to other observations. For an instant, I thought I caught a glimpse of a metallic look about the fuselage, and this really made me feel that the "thing" was close! (Later, John said that he saw this "metallic look" too.) It seemed as though a faint light reflected on the fuselage. (Perhaps the faint light came from the lights of the city or from the lights of the "thing.")

It appeared to me that the fuselage was constructed of many pieces of flat sheets of metal-like material with a "riveted together look." It occurred to me that the fuselage was not of smooth contour. The many "windows" seemed to be lit up from the inside of the fuselage with light that was quite bright. This light seemed to be about the same color as light coming from the windows of our homes. I did not observe anything other than the light in the windows. (It occurred to me that I might see objects or persons, but there was little time for a *good* look.)

My rough estimate is that two-thirds or three-quarters of the fuselage near the front end had windows that were lit up. About one-third or one-quarter of the fuselage toward the rear end was dark or without lights. I did not observe any blinking lights on the "thing" like we have on our planes. From out of the back end of the fuselage came a wide (roughly about the width of the fuselage) long, reddish-orangish-yellowish stream of dusty fire. It seemed as though particles of dust were on fire. These tiny sparkles seemed to make up the tail and the light from it seemed of quite low intensity when compared to the light emitted from the "windows."

I listened intently for some sound from the "thing," but I didn't hear a whisper of a sound! This was the most eerie part of my whole experience! Certainly, there should be some sound from an aircraft that looks so near! It flashed in my mind that perhaps the sound was yet to follow.

I was impressed with what looked to me like low altitude of the craft at this point of my sighting—I thought, around 1,000 feet or less. Also, when the craft was flying near us, it did seem to travel in a flat trajectory. I toyed with the idea that it even slowed down somewhat, for how else could we observe so much detail in a mere flash across the sky? (John doesn't think it slowed down.)

The craft was headed away from us now. I concentrated on the "trail of fiery particles" that seemed to come from the end of the fuselage. I was expecting to see a bright ball of fire close to the fuselage end, but I saw no bright ball of fire. However, I noticed that the trail's light intensity did increase somewhat. Mayor noticed this increase in tail brightness too) but this was understandable, since we were looking at a denser view of "fiery particles." In other words, along the length of the trail instead of the previous width of the trail. Because this light pattern of the craft was at a slight angle from where we were standing, it was possible for a brief moment, to see near the "fiery trail's end" one or a few lit-up "windows," simultaneously.

Upon this observation, I concluded that there must be an outward bulge in the fuselage, especially after taking into account that there were no windows toward the rear end. Also, the simultaneous view of nearly full "trail light" and one or a few window lights gave me the opportunity to compare light intensities again. The light from the window or windows seemed brighter than the trail's light.

All too soon, the "thing" was flying away, low over the treetops toward the Northeast. I could see only the "orangish-colored" light of the trail now. Certainly, SOUND would come from this craft!!! The three of us remained quiet while looking and listening. I was still expecting to hear noise, but, instead, there remained only silence! The three of us remained quiet for awhile, even after the craft was well out of sight. We were all baffled by that.

Then—HULLABALOO!!—we all started talking at the same time! In the course of our expression, the conversation went something like this.
Someone: "It didn't make any sound!"
Mayor: "That wasn't a meteor, because a meteor doesn't have windows, but I'll be damned if I'll report it!"
Marie: "I'm not going to report it either!"
John: "I'm not going to report it!"
(Laughter.)
Someone: "Tremendous speed!"
Marie: "What time is it?"
Mayor, looking at his watch: "Quarter to nine."

Marie: "How high would you guess it was?"
Mayor: "Not more than 2,000 or 5,000 feet, or maybe lower. That thing was really low!"
Marie: "What direction?"
Mayor: "Southeast to Northeast."
John: "Whatever it was, it will be in the papers tomorrow."
As we excitedly compared notes, we agreed with each other on most of our observations. This "agreement" seemed to comfort me, for I certainly didn't want to think that I had just experienced my first hallucination while I was wide awake! All three of us agreed that we had seen something other than any planes we had seen or read about from our Earth. We thought we had seen a "craft of top secret category from our Earth," or that we had seen a "craft from Outer Space."

It was chilly outside and mayor wasn't wearing a coat. Besides, our short discussion of the event seemed to suffice. It didn't seem strange that, so soon after, all three of us went back to our routine of daily living, for, after all, can anything really surprise us in these days in this scientific era?!

A sketch of a zeppelinlike craft, containing ten windows, accompanied this report, with the comment:

I was more interested in looking *into* these windows than I was in studying window shapes. However, I feel strongly that the windows had definite symmetrical shapes, were clearly outlined as the craft passed by, and were lined up in a row, horizontally. I feel safe to stress that the windows did not look blurred or fuzzy, but had clear, definite shapes. I observed, also, that the windows looked quite large. I would say larger than the windows we have in our planes.

A letter from Marie accompanied the foregoing report and referred to the newspaper accounts of the event, dated some sixteen days after the sighting. This letter contained the information that Marie had had two predinner cocktails of bourbon, ice, and water. She had had two glasses of wine at dinner, an after-dinner drink of Irish Mist, and a final after-dinner drink of some other liqueur that "was tasty, not potent, and definitely was not absinthe." Marie attested: "I felt mentally and physically alert by 8:45 P.M.!!!" The letter continues for more than two pages, indicating her woman's activities, and giving character references accompanied by various printed records indicating the high regard the community held for the mayor and for her husband John.

Marie was not the only one who gave a vivid report. Elizabeth, a Ph.D. from Ohio and a teacher of general science, also made a report. She had served in the U.S. Navy during World War II. She claimed that she was very much interested in UFO's. Her report begins by stating: "This is no natural phenomenon. It's really a UFO."

Elizabeth made sketches of the object, which she viewed through field glasses as well as with the naked eye. At first it appeared to her like a meteor or comet. She concluded, however, that it could not have been a "falling star" because of the peculiar behavior and the colors. The object slowed down as it approached the horizon then suddenly became three. The colors ranged orange-white-red-orange, similar to the color of the sun. The objects flew in perfect military formation. The object was flat and the bottom part had a protrusion; it moved very slowly in the NNE direction.

Elizabeth flashed a flashlight in Morse code, SOS, four times. There was no visible response. No noise was audible to the human ear. However her dog, a Boston terrier aged one year five months, who hates the cold, crawled between two trash cans beside the garage and whimpered and lay on the drive between the cans as though she were frightened to death.

Elizabeth reported an effect upon herself, as well:

> After I came into the house I had an overpowering drive to sleep and since I was expecting a phone call at 10:20–10:25 I had to force myself to stay awake. I opened the windows wide in hopes the cold room would help, but even then I dropped off several times. This is extremely unusual behavior for me. I had slept ten hours the night before and had an hour's nap in the afternoon. I had been outside in the cold and should have been wide awake. I felt physically depleted and just had to sleep. This gradually wore off until by 11:00 P.M. [I] was wide awake again. My friend recalled that this had happened to me in 1966 when I saw a UFO then. I had forgotten until she remarked that it had happened to me previously. I did not know others had seen this until I heard about it on the news.

Another report, this time from Indiana, was equally graphic.

About 9:45 P.M. I looked out the window and saw some kind of fire-colored object fly across the valley. About two or three minutes later my cousin, aunt, and my uncle came running into the house yelling and trying to tell me about the UFO they saw. They and some neighbors all observed it from horizon to horizon, which took a very short time.

The object flew at about tree-top level and was seen very clearly since it was just a few yards away. All of the observers saw a long jet airplane, looking like a vehicle without wings. It was on fire both in front and behind. All the observers observed many windows in the UFO. My cousin said "If there had been anybody in the UFO near the windows, I would have seen them."

The next morning we heard it was supposed to have been a meteor. But the other observers and myself know the UFO could not have been a meteor because meteors don't have windows and turn corners like IT did. And it didn't make any noise whatsoever. I believe what we saw was a Flying Saucer. . . .

One observer called attention to the large number of grass fires in the neighboring country on March 4. He wrote: "I do not know if this is true, but I heard there were 72 grass fires in this area on the day following the sighting. I would think there might be a possible connection."

He concluded: "Please send me information on what to do in the event of future sightings. I have often wondered about reports of landings and why people did not shoot it or attempt to capture it or something. I think some effort should be made on the part of the sighter to obtain proof and identification, since he is the only one on hand." Hundreds of people called in to local airports, local police stations, and other authorities who might be able to furnish information about the unusual sighting. These descriptions, reports, and conclusions are all lost forever. But the foregoing selection from the Air Force files clearly demonstrates that something unusual occurred on the night of March 3, 1968, and that it was observed over a wide area ranging from Tennessee and Kentucky in the south to Massachusetts in the north.

What, indeed, caused this remarkable apparition? Earlier that day, Moscow announced that they had placed one of their artificial satellites, Zond IV, in a parking orbit around the earth. Presumably they would reignite the rocket engines later, to send the experimental vehicle into space, for various scientific purposes. Something went wrong with the experiment, however, and the satellite did not achieve as great an altitude as had been planned. In consequence, it re-entered the earth's upper atmosphere, where friction heated the satellite to incandescence and broke it up into several fragments, which gave a spectacular display. The object was at least seventy-five miles above the earth's surface —not, as various observers reported, just over the tree tops or at altitudes of a thousand feet or so. Irregularities and the illumination undoubtedly gave the illusion of "windows," but the satellite did not contain windows. And thus the mystery was solved, promptly and conclusively. The UFO turned out to be a phenomenon that is becoming quite familiar in the space age, the fiery re-entry of a satellite.

Of significance is the tremendous variance of the reports. People are simply not good observers or good reporters of what they see. Hence, when the UFO is a report by only one or two observers, in the Air Force files, how can we reassure ourselves about the reliability of the reporting? Our friend Marie had an impeccable reputation. She, John, and the mayor, were certainly not making things up. This is what they saw. And, if the mystery had not been solved, no amount of reinterviewing these witnesses, by Dr. McDonald or anyone else, could possibly get them to change their conclusions about the character of the UFO. This story carries its own warning. No matter how reliable the observer may seem to be, his estimates of size, shape, appearance, brightness, and other physical characteristics are often very far from the truth.

### 4. The Phantom Plane, Colorado Springs, May 13, 1967

This case is particularly interesting because it consists wholly of a radar observation. There were no visual observations. Robert Low of the University of Colorado Project referred the case to me for comment and analysis shortly after its occurrence. The details are given in the Condon Report (pp. 170–171). An airport radar picked up an image of a Braniff plane, a 720, when it was about four miles away from the field. A second plane, a Continental Viscount, also appeared on the screen, but it does not figure in the case.

Just as the radar operator registered the Braniff plane on the screen, he detected an extremely faint target about two miles behind the 720.

Further observation disclosed that the radar UFO was following and gradually overtaking the 720. Several alerted observers watched visually for the unknown plane, but they saw nothing. They were afraid it might collide with the Braniff plane. Finally the 720 came in and landed. The radar UFO vanished from the screen, and nothing was ever seen of it. What was this ghostly UFO?

Having encountered similar cases when I was serving in the Navy, advising on problems of radar propagation during World War II, I sent in my analysis of the facts that were presented to me at the time. Here was a UFO moving approximately twice as fast as the main target, gradually catching up with it. I could account for this phenomenon in only one way. The pulse of radar reflected from the 720 came to earth somewhere within a mile or so of the airport, where something reflected it directly back to the same aircraft, which reflected again the very weakened signal. The second echo, traversing about twice the path length of the first, would always appear to be about twice as far away as the primary target. And, of course, as the 720 landed, the UFO would vanish from the screen.

The foregoing explanation requires that we account for the reflector that captured the downcoming radar pulse and directed it back almost precisely to the target plane. There is a device well known in radar as in optical work as a "corner reflector." Such a device consists of three reflecting surfaces, at right angles to one another, such as the inside walls and floor of a room. A rubber ball tossed into such a corner will bounce successively against each of the three surfaces and then return practically along its original path, except for the fact that neither the surfaces not the ball are completely elastic. But a light beam from a searchlight reflected against three mirrors set at right angles will come precisely back on its initial path. And so will a radar wave from any metallic corner reflector.

There are many natural reflectors available. The inside of a dump truck, the corners of a metal fence or building, the inside of an empty, open freight car, for example. No matter what the orientation was initially, such a corner reflector would intercept the radar pulse and direct it back toward the 720. This simple explanation will account for most of the major facts of the phantom UFO. Mr. Low accepted it at the time, and the Air Force Project Bluebook also acknowledged it as correct.

To my surprise, the Condon Report failed to accept my explanation and lists the Colorado Springs sighting as an "unknown." They have done so, it appears, on the basis of information furnished by the observers more than a year after the event occurred. They questioned whether the UFO actually overtook the 720. The rate of approach during the final seconds, someone suggested, did not completely accord with my hypothesis of a phantom reflection.

We are not told who conducted the later survey. There is plenty of room for slight variations on my original hypothesis. Corner reflectors are common and there might well have been more than one responsible, especially as the plane came in toward the field, where many metal buildings or other structures may have been present. A few of the UFOlogists conclude that the failure to see a phantom radar UFO was in itself absolute proof that an invisible, unidentified plane was pursuing the 720. This is absolute nonsense! We had many similar cases of phantom UFO's all over the world during World War II. We found the proper explanations, trained our radar operators to recognize them, and thereafter did not worry unduly about them. I have earlier referred to a related but much more spectacular phenomenon known as the "galloping ghosts of Nansei Shoto."

### 5. Do Flying Saucers Move in Straight Lines?

Aimé Michel, in his book *Flying Saucers and the Straight-Line Mystery* (New York: S. G. Phillips, 1958), makes the following suggestion: Take all of the saucer sightings for a given day, from midnight to midnight. Plot on a map the places from which the sightings were made. You will then find that they tend, within a reasonable margin of error, to fall along straight lines. Michel terms this property of the flying saucers "orthoteny."

I found Michel's original book, in French, unconvincing. I could see no reason why the places of observation rather than the probable locations of the saucers should have special significance. And his mathematical discussion was full of obvious errors. I didn't even consider his hypothesis worth mentioning in *The World of Flying Saucers* (co-authored by Lyle G. Boyd, New York: Doubleday, 1963). However, I finally decided to examine Michel's extravagant claims. I also studied the U.S. version of his book with the supplement by A. Mebane, which deals with the great French saucer "flap" of 1954.

Here is the basic problem: Michel arbitrarily defines a "day" as the interval from one midnight to the next. He collects all the French sightings he can find for that interval, mostly from newspapers. He then plots them on a map of France. Presumably he knows or can find out where the observer stood when he saw the UFO. But clearly he cannot expect to know above what spot the saucer was even if he knew its direction from the observer. It could be a large, bright, distant UFO or a small, faint one nearby. So Michel decided he had to be content with the observer's location as a fundamental statistic, even though the UFO, if it had an altitude of 30,000 feet and appeared 30 degrees above the horizon, could be ten miles from the point of observation. This fact alone injects considerable ambiguity into Michel's analysis if it does not make his conclusions highly questionable.

Michel found, if he plotted the sightings made and reported during a single day on a map of France, he could draw straight lines on this map in such a way as to connect three or more points. Of course, since any two points must define a straight line it is necessary for at least three points to be colinear if the alignment is to have any significance. Michel decided that the number of such straight lines he could draw on the map was far greater than one would expect on the basis of chance. He therefore concluded that orthoteny was something real, a phenomenon proving in turn that UFO's themselves were real. Then Michel noted something else. The geometric pattern of the lines formed a complex sort of a star. He regarded this feature as also related to the peculiar orothotenic pattern of saucer movements.

Michel does not give us too clear a picture of his methods of data selection. He used newspaper accounts for the most part. But we do not know how representative his sample is of French newspapers. From time to time, whenever it suited his purpose, he brought in a foreign sighting. Such practice is forbidden by proper statistics, since he does not include a representative sample of foreign sightings. We must therefore disregard, from the outset, all but the French sightings. At the very least, before considering sightings elsewhere, a proper statistical background must be laid.

Michel chose to give equal weight to all French sightings, good or bad. By including the poor sightings he greatly increased the number of sightings available for analysis. One must therefore be particularly alert to accidental alignments caused by the number of sightings. The second arbitrary decision that Michel made was his subdivision of the time, choosing to regard as a unit all the sightings that occurred during a given day from midnight to midnight. I am not questioning that decision; he may make any rule he chooses in advance, so long as he never departs from it.

The mathematical expression for the number of straight lines expected to occur by chance is formally correct in the section written by A. D. Mebane. However the mathematical formula given by Michel himself was completely wrong. If three (or more) points approximately line up, what departures from exact linearity will one permit, while calling the line "straight"?

One long chapter of Michel's book, written by Mebane, deals with U.S. sightings. The most one can say for this section is that the statistical formulas are correct. But the author wants to believe in saucers and finally argues against the validity of his own statistics.

The derivation of Mebane's formula is much simpler than he makes it. Suppose, for example, that we have a map containing six observations of saucer sightings. And suppose we want to predict the number of times that three observations lie along the same straight line. Label the six observations: *a, b, c, d, e, f* and combine them in groups of three, as follows: *acb, abd, abe, abf, acd, ace, acf, ade, adf, aef, bcd, bce, bcf, bde, bdf, bef, cde, cdf, cef, bef.* There are twenty combinations of six different things taken three at a time.

One easily proves that the general formula for the number of combinations of $n$ things taken $m$ at a time, abbreviated $\binom{n}{m}$, is

$$\binom{n}{m} = \frac{n!}{m!\ (n-m)!} \tag{1}$$

where the exclamation point signifies what the mathematicians call "factorial," the product of all the integers from 1 to $n$:

$$6! = 1 \times 2 \times 3 \times 4 \times 5 \times 6$$
$$3! = 1 \times 2 \times 3$$

and so on. When 0! happens to occur, its value is 1. Thus, the number of combinations of six things taken six at a time is:

$$\binom{6}{6} = \frac{6!}{6!\ 0!} = 1 \qquad (2)$$

and as in the example of six things taken three at a time,

$$\binom{6}{3} = \frac{6!}{3!\ 3!} = \frac{1 \times 2 \times 3 \times 4 \times 5 \times 6}{1 \times 2 \times 3 \times 1 \times 2 \times 3} = 20$$

On our map of six observations we can draw twenty zig-zag lines connecting three points. Whether or not any given zig-zag line is straight depends on our definition of straight. Let us try to define it. Connect any two of the points—preferably the ones farthest apart—by a straight line. Then draw, parallel to this line, two other straight lines, two and one-half miles on either side of the original line. These two lines define, with the boundary of the map, a roughly rectangular corridor five miles wide, running across the map. If the third point falls in this corridor we shall say that the line is "straight" with a small allowable margin of error. I have suggested that it be five miles wide, because Michel picks that figure. However, on a number of Michel's maps, the width often reaches and occasionally exceeds ten miles. Let us defer this question momentarily.

Suppose that this corridor occupies a fraction, $f$, of the entire map. We have used up two of our points to define this rectangle. If we are studying three-point lines, the probability that the third point will lie within the rectangle is $f$. If we are counting four-point lines, the probability of getting the two extra points into the corridor is $f \times f = f^2$. In general, for $m$ points, the probability is $f^{m-2}$, or $f$ multiplied by itself $m-2$ times. And so, multiplying this value by the number of $m$-point lines, we get the probable number of $m$-point lines, from $n$ observations

$$N = f^{m-2}\binom{n}{m} = f^{m-2}\frac{n!}{m!\ (n - m!)}$$

This formula agrees with the one in Michel's book, given by Mebane, but my derivation of it is simpler. Mebane does not properly define the corridor.

Now consider one of Michel's prize examples, October 7, 1954, for which he plotted 27 sightings. Michel marvels at finding 19 three-point lines. As a matter of fact Mebane's formula predicts that no less than 37 three-point lines should occur by pure chance. Where then, are the missing 18, which added to Michel's 19 go to make up the theoretical 37? No one has suggested one obvious interpretation, just the reverse of Michel's, that the saucers have moved in such a way as to *avoid* straight lines. Actually, the answer is much simpler. Michel simply failed to draw other three-point straight lines quite as well as those he delineated originally. He had failed to notice many three-point alignments. As for four-point straight lines, Michel drew 3; theory predicts 2.7. Pretty good agreement!

At this point, Mebane, who seems about to dispose logically of Michel's straight lines, produces a red herring. He scatters 27 catnip seeds on a map and finds that the number of three-point and four-point lines drawn on this admittedly random pattern agrees pretty well with theory. The fishy part? Mebane suddenly abandons his statistics and notes that the catnip patterns are more jagged and less "boxed in" than those drawn by Michel. He substitutes a subjective test for his mathematical deductions, suggesting that Michel's lines represent real orthoteny whereas the catnip seeds are only pseudo-orthoteny. This is patent nonsense. Had he tossed the seeds on a map of France, some would have fallen beyond the borders, in Spain, Germany, Belgium, or the ocean. Of course Michel's patterns are boxed in. The points were confined to France alone. Clearly the lines must fall into some sort of pattern. But trying to read something into the figure is a little like attaching significance to the changing form of a fleecy cloud on a summer day.

In short, the statistical analysis has revealed the three-point and four-point lines as accidental features. More than that. On some of Michel's diagrams, where the number of three-point lines is appreciably less than that indicated by chance, a re-examination reveals that he missed drawing in a goodly number, whether by accident or design we do not know.

Michel has discussed two other features purporting to prove the reality of his three-point lines, on which his original argument largely depended. The first of these relates to the number of intersections at a common point, suggestive of a central control, directing the saucers along lines radiating outward from a point. The spiderweb pattern of Michel's map is his best example.

Michel introduces another cute gimmick. He takes his maps in pairs, superposes them, slides them a trifle and rotates them. In a few instances, he finds a similarity of pattern, and suggests that the saucer operators had kept the same basic flight pattern but had rotated it through some angle on the second night.

First of all, the pattern depends almost wholly on those questionable three-point lines. Second, with enough maps, both slid and rotated, it takes only a little imagination to see occasional similarities. But third and most important of all, the continued testing of hypothesis after hypothesis as to what the patterns might signify in itself reduces the chance that any discovered correspondence represents something real. This is an old trap that even experienced statisticians have occasionally fallen into.

What, then, of Michel's great prize, on September 24, 1954, when six out of nine sightings lay on a line drawn from Bayonne to Vichy (hereafter called the Bavic line)? A simple application of the formula indicates that the odds against such an alignment are about 5,000,000 to 1. However, this figure applies only if the statistics are carried out by the original rules.

First of all, Michel notes that a sighting made on September 24, near Vierzon, at 3 A.M., lay near another line drawn for the previous day. So Michel deftly reports it as September 23–24 and plots it on the map for September 23. This trick neatly gummed up the statistics for September 24. Clearly Michel's prize should have been six out of ten instead of six out of nine. This reduces the probability to 2,000,000 to 1. Now that Michel has called our attention to his careless procedures, we note that two of his reported six sightings occurred, as he states, "about 11 P.M." How accurate is the time determination? If they occurred after midnight one must remove them from the map! Or, if time is so significant, why did Michel originally divide the day exactly at midnight? Do the saucers operate on local French time? Michel himself evidently has doubts and so again plots two observations for October 4 on the map for October 3, because they seemed to fit better with his preconceived ideas. And then, just to be safe, he plots them again on the map for October 4, completely disregarding a fundamental rule of statistics that does not allow us to change our mind in the middle of an analysis.

We have every reason here to suspect Michel's methods because he has specifically stated that he was "studying very closely the cases of landings reported along the Bavic line." Some friend sent Michel a Portuguese sighting for September 24, 1954. Michel claims that he was "upset" by the sighting until he found to his "amazement" that it fell on the extended Bavic line. It is highly probable that the correspondent sent Michel the sighting because it lay on that line. If the sighting had *not* fallen on the line, rest assured we should never have heard of it. Anyway, statisticians should never get upset. They must accept whatever the analysis shows. The Portuguese sighting is irrelevant since the statistical data for all of Portugal were not included for that date.

Michel, becoming convinced of the reality of the Bavic line, now searches for other sightings along the line and turns up two more, sightings of several years later at Tulle and Brive. He asks "Could that be a coincidence?" Evidently he expected a resounding negative reply!

Remember, he was limiting his search to the narrow rectangle along the line. But during the three-plus years between the two sightings, how many saucer reports had come in from all over France? The two extra sightings, found in this manner, are indeed mere chance and subtract from rather than add to the validity of the Bavic line.

The statistics are somewhat confused, because the correct procedure applies to only *random* sightings. A search along the line does not reveal the random character necessary for statistics to apply. The two added observations have the same effect as if they were outside the corridor. Here the implication is that if one wrote to the same number of places outside the corridor, he would have received an equal number of new reports. Thus, the new probability is

$$\left(\frac{1}{40}\right)^2 \binom{10}{4} = \left(\frac{1}{40}\right)^2 \frac{10!}{4!\ 6!} = \frac{21}{160},$$

or about 1 in 8. Flying saucer enthusiasts will probably violently reject my claim that two discoveries *along* the line should *reduce* rather than increase the probability that the line is real. Such are the facts of life. The reason is simple. If someone can write to a few towns in the corridor and get even two affirmative replies, the chances are that a similar questioning of towns outside the corridor would have produced at least the same number of affirmative replies. One might even argue that, since the area outside is forty times larger than that of the corridor, one should multiply the number of saucers reported by 40.

In the same way Michel adds a sighting at Vauriat on August 29, 1962. He selects it as "the most sensational French sighting of the year." What a change of statistical method! Originally he included all sightings, good or bad. Now he selects one sighting by a subjective procedure. Whatever merit the original Bavic line may have had back in 1954 has completely vanished.

What more can we learn about Michel's methods? From time to time he brings in a sighting from Rome or Africa, *if* it fits with his pattern. Clearly he disregards a sighting if it does not fit. This procedure is against all the rules of statistics unless the researcher had originally planned systematically to secure all the data over a larger territory and had included all of the observations in his analysis.

I find quite revealing Michel's statement that October 12, 1954, marked the crest of the wave of French sightings. "Unfortunately, witnesses and reporters alike were getting far beyond their depth, and only a limited and inadequate number of all these sightings were dated with any exactness." He complains of the difficulty of dealing with such statistical material. If he had stuck to principle and refused further analysis I should have applauded. But he nonetheless gives maps for another week, containing a bare skeleton of sightings. Straight lines appear, of course, but one does not know how to evaluate them since Michel does not reveal his criterion for the rejection of the much more numerous nonlinear observations.

Because the Bavic line, extended around the earth, runs by chance through Brazil, Argentina, New Guinea, and New Zealand, among other countries, Michel finally unveils his sweeping conclusion: The Bavic line possesses planetary significance. He drops all pretense of using statistics, which he never employed correctly anyway. He makes short shrift of the number of U.S. sightings, attributing them to the launching of the first Sputnik.

One might excuse Michel if he used the available Argentine sightings, for example, to investigate the multiple line-ups in that large South American country. It might have been convincing if such a study, made without reference to the French sightings, established a line that seemed to be an extension of the Bavic line, within reasonable error. But all that Michel succeeds in proving is that a line, drawn from the middle of France to the middle of Argentina, passes through Brazil and some other countries where flying saucers have been reported.

Jacques Vallee has presented formulas for calculating the great circles supposed to represent global orthoteny. Let $T$ be the longitude at which the great circle intersects the equator and $u$ the inclination of that circle to the equator. Then a point on that circle, with longitude $L_i$ and latitude $\Phi_i$, conforms to the equation

$$\sin (T - L_i) = \cot u \tan \Phi_i.$$

This equation has two unknowns, $T$ and $u$. Hence two points giving two equations serve to determine the great circle.

If instead of 2 we have $N$ points lying on or close to the line, the extra $N-2$ equations are redundant. Alternatively, we may find some way of averaging the points to derive the best possible great circle. Vallee applies the method of least squares for this purpose. He makes the following substitutions:

$$x_i = \frac{\tan \Phi_i}{\cos L_i} \text{ and } y_i = \tan L_i$$

$$A = \frac{\cot u}{\cos T} \text{ and } B = \cos T.$$

Then equation (1) becomes:

$$y_i - Ax_i - B = \varepsilon_i$$

where $\varepsilon_i$ is the error if $x_i$ and $y_i$ do not lie exactly on the great circle.

This formula, however, gives artificially high weight to points near $L_i = 90°$ for which both the tangent and the reciprocal of the cosine go to infinity. I am sure Vallee did not intend to give undue weight to the U.S. observations. His equations also give undue weight to observations from high latitudes.

To avoid both pitfalls, I should proceed as follows. Let

$$\sin \Phi_i = a_i; \cos \Phi_i \cos L_i = b_i; \cos \Phi_i \sin L_i = c_i$$
$$-\sin T \tan u = X; \cos T \tan u = Y.$$

Then, the equivalent of (2) is:

$$\varepsilon_i = a_i + b_i X + c_i Y,$$

and the sum of the squares of the errors becomes

$$S = \sum_{i=1}^{N} \varepsilon_i^2 = \sum_{i=1}^{N} (a_i + b_i X + c_i Y)^2.$$

Differentiating to get the minimum of $S$, we have

$$\left.\begin{aligned} \frac{\partial S}{\partial X} &= 2 \sum (a_i + b_i X + c_i Y)\, b_i = 0 \\ \frac{\partial S}{\partial Y} &= 2 \sum (a_i + b_i X + c_i Y)\, c_i = 0 \end{aligned}\right\}$$

We thus get two simultaneous equations to solve for $X$ and $Y$, as follows:

$$\left.\begin{aligned} X \sum b_i^2 + Y \sum b_i c_i + \sum a_i b_i &= 0 \\ X \sum b_i c_i + Y \sum c_i^2 + \sum a_i c_i &= 0 \end{aligned}\right\}$$

These equations apply for any value of the latitude or longitude. A slightly different set will be necessary when the inclination is nearly 90°. These equations are certainly preferable to those given by Vallee. However, the applicability of least squares to the problem is somewhat doubtful. For least squares to work, the errors, $\varepsilon_i$ , must be truly random. We have no assurance that this is so. For example, a random distribution would result if we used the line as a target and established the stations by throwing a dart. Nevertheless, as I have previously noted, the global orthotenists will get the shock of their lives when they use these equations in a truly global sense.

For a short arc like the Bavic line the equations are not sensibly different. But Michel has claimed that certain sightings in Brazil, Argentina, New Guinea, and elsewhere are extensions of the Bavic line. I predict that the errors will be enormous when one tries to put a great circle through all the sightings.

Vallee further states that the distances between selected stations, divided by selected integers, give approximately the same figure. This new claim, in my opinion, is no more convincing than the other orthoteny "proofs."

Experienced statisticians well know that, when a person starts to search for such relations, he can always find them even in a series of purely random numbers. The streets of Las Vegas and Monte Carlo are paved with the hopes of gamblers who have had similar illusions.

The foregoing analysis contains only a fraction of my scientific objections to Michel's claims for orthoteny. He proposes other relationships such as persistence of patterns from night to night. He further asserts that the multiple intersections of straight lines have special significance. None of these claims holds up under critical analysis. The straight lines and all their associated properties are clearly the result of extremely bad statistics. The claims for high precision are invalid. The lines exist only in the imagination of Michel and his followers. Orthoteny turns out to be just one more of the flying saucer myths.

I should like to draw attention to a singular coincidence closely related to orthoteny, but one that has not been previously noted, as far as I can determine. In one of the famous Paris book stalls along the bank of the Seine, I came across a book of science fiction entitled *L' Exile de l'espace: Adventures dans le système solaire* ("The Exile of Space: Adventures in the Solar System"), by Pierre Devaux, published by Editions Magnard, Paris in 1948. Chapter 3 of this book, written years prior to Michel's volume, contains a remarkable description of a phenomenon similar to orthoteny.

An academician, lecturing to his colleagues, apprises them of a vast network covering all of western Europe with a strange cobweb dating from prehistoric times. The points forming the basis of the network, he claimed, were all the locations of villages having names derived from Alesia. He listed a number of places whose names were clearly similar, such as Alise, Alaise, Aizieux, Alyes, Alliezes, les Allys, les Alyscamps de Arles. He continued with other less obvious forms like Calais,

Cles, Calaize, Calice, and Versailles. He further pointed out that France, today, has some thirty-three villages named Versailles.

He continued: "Cast your eyes on this map. These rigorously straight lines have been traced to join the localities carrying the names, more or less deformed, of Alesia. Note how this line drawn from Spain to Poland starts from Aliseda, crosses the Pyrenees, passes Aizieux l'Allée, Eauze, cuts through Cales, Calais, Ales, pierces Allis near Rocamadour, Calais Puy-de-Dome, Montalays, Alaise du Doubs—finally finishing up at Kaliszin Poland" (translated by the author).

He points out other alignments, some continuing into Italy, and then makes the point: "These lines are straight. *Inexplicably straight,* especially when one thinks of them being traced by prehistoric man. I have faith that one cannot speak of this as a coincidence, that a straight line could not possibly be drawn on a map of Europe in such a fashion as to traverse more than three villages, four at a maximum—villages all bearing the name. These lines, you must admit, diverge in such a manner as to form a gigantic European star centered around this precise point, Alaise, situated on a tragic plateau 18 kilometers south of Besançon, *Alesia Mandubioum,* the unique, the true Alesia, which must have been the point of departure for the return of the Gauls."

The remainder of the story is mostly inconsequential. It deals with the solution of a Caesarian cryptogram, purporting to establish the original thesis of the reality of the network, in terms of prehistoric towns named Alesia. It contains arguments for the significance of patterns within patterns, supposedly confirming the alignments, with the final "identification" of d'Ys as Ulysses and Paris as Al-uzza (Venus). All good fun and good science fiction. Highly reminiscent of and about as authoritative as orthoteny.

In passing, it is interesting to speculate that Michel may have read this book and was induced thereby to develop the orthotenic hypothesis. The correspondence, even to the star-shaped outlines, is otherwise a remarkable coincidence.

### 6. UFO's in Art

The realization that famous early artists had dealt with the UFO phenomenon suddenly struck me in the late summer of 1958. I was passing through Brussels, where the World's Fair was then in full swing. As a side line, I visited the local Museum of Ancient Art—a marvelous collection in itself, well worth the trip to Brussels. There I came across a remarkable painting, which had a number of queer objects flying through the sky. One of these was clearly a saucer with outstretched wings, manned by a creature with an egg-shaped body and a head obscured by a hat resembling an inverted lamp shade. I was almost struck dumb! Here was the prototype of a flying saucer! The picture was called *The Temptations of Saint Anthony the Priest.*

The artist was the famous Flemish painter Hieronymus Bosch (1460–1526). He might be best described as the founder of a school of artists devoted to portraying the fantastic. The more famous of his disciples included Lucas Cranach the Elder (1472–1553), Pieter Breughel the Elder (1525–1569), David Teniers the Elder (1582–1649), and the French engraver Jacques Callot (1592–1635). All of these men were satirists of a sort, combining humor and beauty in their art. And of all these, I consider Bosch the greatest. He followed Dante by about a century and a half and was clearly influenced by Dante's concept of the universe and the then popular religious concept of a heaven beyond perfection and a hell filled with torment and suffering.

Some say that Bosch was mad, and perhaps he was. His paintings clearly reveal his obsession with moral problems, within the areas of sin, damnation, and salvation. His paintings *The Garden of Earthly Delights, The Seven Cardinal Sins,* and *The Haywain,* all deal with these questions, vividly, beautifully, and sometimes terrifyingly. These three pictures hang in the Prado Museum, in Madrid, which contains the greatest collection of Bosch paintings in the world. The Spanish fondly refer to the artist as "El Bosco."

Bosch's fantasy reaches its ultimate in *The Temptations of Saint Anthony.* According to early legend, in about 300 A.D. the priest became an ascetic and retired from all worldly things to live as a hermit in the Egyptian desert. He is regarded as the founder of Christian monasticism. The story goes that Satan tried everything to dissuade the Saint from his mission. He sent beautiful women to beguile him and hordes of demons to frighten or torture him. The holy man resisted all of these pressures and remained calm, while continuing to read the scriptures.

The picture is a triptych, consisting of a main picture with two hinged panels on either side, which can be folded to hide the painting completely. On the left panel several companions support the weakened saint, while weird creatures wander around below and above, flying in different kinds of vehicles. One of these appears to be a sort of ceramic casserole with handle. The right panel depicts a naked woman trying to seduce the saint, while two creatures fly overhead on an enormous winged fish.

The central panel is the most spectacular, with the fires of hell burning brightly in the upper left. The little man in the flying saucer, which had originally caught my eye, is partially enveloped in the dense smoke above the fires. What could easily pass as the luminous train of a modern rocket ship from some vehicle appears in the left-hand corner. A swan boat engages another craft in heavenly jousting. A demon, carrying a ladder, flies in the midst of the conflagration. An owl perches on the head of a man with a bulbous nose and the face of a pig. A creature whose head consists of a horse's skull plays a decrepit harp with iron-tipped fingernails. A woman, formed from a decayed hollow tree, sits side-saddle on a giant rat while she holds a child wrapped like a mummy. I could mention hundreds of other details, for the picture is highly complex. Through all this turmoil calmly sits the saint, demonstrating the power of good over evil. By all means see this picture if you should go to Brussels! A duplicate, also by Bosch, hangs in the Lisbon Gallery.

The painting tells us little about flying saucers, except that the concept is very old. But it does indicate the tendency of people to interpret natural phenomena in terms of the philosophy of their times. Devils and demons were very real to those who lived in the time of Bosch.

The temptations of Saint Anthony proved to be a popular subject with many painters who followed Bosch. The Prado has at least one by Teniers. Peter Paul Rubens chose it for one of his paintings, but characteristically he emphasises the seductive women rather than the demons. Callot's most famous etching deals with the same subject. I was fortunate enough to acquire one of his priceless originals in a junk shop, in Cahors, France, for a paltry $5.00. It has many of the elements of Bosch's representation. If anything, the demons are more numerous, more active, and more mischievous in the Callot picture. But they have the same out-of-this world look, spouts for noses and caricatures for faces. Satan himself flies overhead, dominating the scene directing the activities of his minions while the saint, oblivious to it all, calmly continues his meditations.

I suppose that the UFOlogists could cite this painting as evidence that flying saucers were known in times long past. But that is precisely my own thesis. UFO's are by no means new. We get only new interpretations of their significance.

## 7. Flying Saucers of the Bible

When, in 1953, I pointed out that flying saucers are mentioned in the Holy Bible, I inadvertently opened up a Pandora's box. Most of the leading writers on UFOlogy got into the act and made similar claims, without credit to me, of course. Their UFO's were manned machines from a super civilization, intruding into the affairs of ignorant men.

I pointed out that two famous visions of the prophet Ezekiel, recounted by him in chapters 1 and 10 of Ezekiel, were in fact singularly accurate descriptions, albeit in symbolic and picturesque language, of a phenomenon well known to meteorologists, technically called "parhelia."

This apparition assumes a variety of forms of which the most common is a ring of light encircling the sun. This is not a rainbow, as some people have mistakenly called it, but an optical effect caused by the passage of sunlight through a thin layer of ice crystals, usually associated with cirrus clouds. Occasionally, two patches of light, sometimes as bright as the sun, occur on one or both sides of the sun at a distance of about 23 degrees. These "sundogs" or "mock suns" tend to be most conspicuous when the sun is low in the sky.

Sometimes a second outer ring appears, enveloping the inner one. A vertical and a horizontal streak of light may cross both rings like the spokes of a wheel. Indeed here is a reasonable and simple explanation for the "wheel in the middle of a wheel" that Ezekiel saw. Although the two wheels are singularly devoid of color, except for a tinge of amber on the inner edge of the smaller wheel, an inverted bow colored like the rainbow with sapphire at the top extends above the outer wheel.

The overall effect of this rare, complete parhelic display is that of a huge chariot, with one difference, as Ezekiel himself noted. As the wheels "were lifted up from the earth" (following the rising sun), "they turned not when they went."

In the early days it was customary to carve the spokes of a wheel to the form of various creatures. It is, therefore, not surprising that Ezekiel visualized living forms in the four bright sundog condensations of the inner ring. The white feathery clouds of the spokes and also of the inner ring suggested wings, two covering the body and two outstretched. The eight outstretched wings of the four creatures formed the inner wheel. And there were "eyes" in the outer wheel, which I take to be the spots of brightness commonly seen in the apparition.

The correspondence, despite the figurative description, can scarcely be accidental. To see whether my identification was original or not, I corresponded with the Vatican expert on Ezekiel, who replied that the idea was new to him but that he raised no objection to my interpretation if I did not imply that Ezekiel's vision was not divinely inspired.

The complete parhelia, with the appearance of a chariot, are rare events. Like Ezekiel, I have seen only two during my lifetime, one in Colorado when I was a boy and the other in Alaska, in 1954. They are indeed spectacular. No wonder that uninformed, credulous people the world over and throughout history have regarded them with superstitious awe, as portents of some dreadful event. I found a record of a woman who had conceived during such an apparition and who nine months later gave birth to quadruplets. In the face of such clear evidence, who could doubt the malevolent influence of such a vision?

The Scriptures carry other references to allied phenomena. Less complete pictures appear in Isaiah 66:15, and Jeremiah 4:13. The fiery chariots mentioned in II Kings, which took Elijah to heaven and shielded Elisha from harm, are probably also related. In Daniel, chapter 7, we find an obvious description of a similar apparition, in which the whole vision was regarded as a multitude of beasts with horns. I should point out that as the sun rises higher in the sky, the parhelia develop curved, hornlike protrusions. In Revelation, the multihorned beasts of the Apocalypse appear at least four times, in chapters 5, 12, 13, and 17. The imagery is clearly related to parhelic phenomena. Moondogs are not uncommon, but fully developed paraselene are very rare, probably because the appendages are so faint.

Having tried my hand in one phase of biblical exegesis, I thought of seeing whether other phenomena reported in the Scriptures might also have an explanation in terms of natural phenomena. I have no intention of irreverence. It is certainly not irreverent to point out that the "bow in the cloud," mentioned in Genesis 9:13, marking the end of the Noachian flood, was indeed a rainbow, sunlight broken up by the droplets of rain in much the same way that fine ice crystals can form parhelia.

In the same spirit, consider Exodus 3:2. "And the angel of the Lord appeared unto him in a flame of fire out of the midst of a bush: and he looked, and, behold, the bush burned with fire, and the bush was not consumed." There is indeed a rare natural phenomenon that can produce such an apparition: a form of lightning discharge. The most familiar form of lightning consists of a bolt that leaps from heaven to earth, usually striking some sharp object like a mountain peak, a church steeple, a chimney, or a lightning rod. On occasion, however, an electrical discharge can occur upward, from the individual branches of a tree or bush, which seems momentarily to be on fire, though it remains undamaged by the experience. It often appeared in the rigging of old sailing ships, whose sailors called it "St. Elmo's Fire," an Italian corruption of St. Erasmus, the patron saint of the Mediterranean. The superstitious sailors regarded the phenomenon as a favorable sign, evidence of the active presence of the saint. The phenomenon is also sometimes called "corposant," a word signifying "holy body."

Many persons have tried to find a natural explanation for the "star in the east," that accompanied the birth of Jesus. A conjunction of four bright planets, which occurred about then, may be the explanation. Others have suggested a bright comet. I am more inclined to accept the often-expressed view that legends about the appearances of bright stars have been attached—after the fact—to the births of many famous figures. But here, except for the fact that Venus has often been reported as a UFO, we are wandering from the primary field.

The vision of Jacob, Genesis 28:12, may also be a UFO phenomenon. "And he dreamed, and behold a ladder set up on the earth, and the top of it reached to heaven: and behold the angels of God ascending and descending on it." There is indeed a possible natural explanation: a full-scale display of the aurora borealis. The beams of ions and electrons, originally ejected from the sun by an explosive outburst, are focussed by the earth's magnetic field. They enter the earth's upper atmosphere and cause the gases to glow. As the aurora increases in brilliance and activity, filling a large part of the heavens, we can look parallel to the magnetic axis, where the charged particles are entering the earth's atmosphere. The effect is one of perspective. The auroral rays diverge from the magnetic axis forming a sort of corona. We seem to be looking through a large, hollow cylinder. The appearance could easily suggest a ladder. And the rapid movement of the light pattern could appear like angels.

There are numerous biblical events for which I can find no possible natural explanation. For example, I do not believe that the sun and moon "stood still in the midst of heaven, and halted not to go down about a whole day" at Joshua's command. For that would require the entire earth to stop rotating. What would have happened to the energy of rotation? I regard the story as symbolic, not factual.

Another account that appears to require a temporary suspension of the laws of nature appears in Exodus, chapter 15, the parting of the waters of the Red Sea, allowing the Israelites to pass through on dry ground. And then the waters returned to entrap the pursuing Egyptians. I have seen explanations attributing the phenomenon to a strong wind that parted the waters. But there is another possible explanation that happens to lie within the field of UFO's. I once observed the phenomenon myself and can testify that the effect is startling—almost frightening. Many years ago, I was standing on a small rise in Death Valley. The day was very hot and the sky deep blue. But the surprising fact was that an enormous lake completely surrounded me, stretching out to the violet mountains of the distant horizon. My hill, which in fact was only a slight elevation above the desert floor, seemed like a tiny island in a vast, blue sea.

I had heard of this phenomenon, of course, and had seen some displays. But never before or since have I seen such a perfect mirage. The "lake" actually consisted of distant light from the sky, light bent upward by a layer of hot air near the ground, so that the sky itself appeared below the mountains. No wonder that thirsty travelers, lost in the desert, can be deceived into thinking that a body of water lies near at hand.

I watched as my companions descended the hill and began to enter the "lake." The shoreline was not sharp. It consisted of horizontal streaks of shimmering silver. The illusion of a man descending into water was almost perfect. The lower part of his body disappeared first. I could even see a "reflection" of his torso in the "lake." Finally only his head was visible and then even that vanished. Had the water been real he would surely have drowned.

When I began to follow, another remarkable thing happened. The shoreline ahead began to recede. My companions reappeared. As I looked back, I could see the "water" closing in behind me. This sort of mirage could have produced the phenomena recorded in Exodus: the parting of the waters and the disappearance of the pursuing Egyptians. Mirage has certainly been responsible for a number of recent UFO's. The hot air over a highway will produce a mirage that looks like a wet spot on the pavement, receding as a car advances. On occasion, this patch of light will assume the shape of a cigar and appear to have a metallic sheen. If a layer of hot air can produce an image of the sky against the ground (an inferior mirage), a layer of cold air can cast an image of something on the ground against the sky just above the horizon (a superior mirage). The UFO files contain many cases that can be fully accounted for by mirage effects.

## NOTES

1. Charles Fort, *Lo* (New York: Claude H. Kendall, 1931).
2. McDonald, "Statement on UFO's" (Tucson: NICAP, 1968).
3. Donald H. Menzel, *Elementary Manual of Radio Propagation* (New York: Prentice-Hall, 1948).
4. *Ibid.*, p. 178.
5. *Ibid.*, p. 204.
6. *Symposium on Unidentified Flying Objects*, Hearings before the House

# 7

## Unusual Radar Echoes

KENNETH R. HARDY

Ever since radar first probed the atmosphere, scientists concerned with the interpretation of the returned signal have been intrigued by mysterious echoes, or "angels," from invisible targets in the apparently clear atmosphere. The nature of these targets as proposed by various investigators falls into four categories: (1) surface and airborne targets below the line of sight which are brought into view by anomalous propagation, (2) insects and birds, (3) direct backscatter from sharp gradients or fluctuations in the index of refraction in the clear air, and (4) unidentified flying objects (UFO's). This chapter outlines some of the key properties of the various types of clear-air and unusual radar echoes and describes briefly how the targets responsible for these echoes can be identified and how they are related to atmospheric structure and processes.

Multiwavelength ultrasensitive radars located at Wallops Island, Virginia, have been used over the past five years to study radar echoes from the clear atmosphere. Such clear-air echoes are detected consistently with these radars. Although the echo sources were difficult to identify initially, there is no longer any mystery about the general mechanisms which give rise to the echoes. In all of the detailed investigations which have been conducted with the Wallops Island radars, *all* classes of targets fall readily into category 1, 2, or 3 listed above. At no time has any object been detected at Wallops Island which remained unexplained and was therefore put in the category of a UFO.

*Anomalous propagation* is the propagation of radio waves in a direction somewhat different from the direction normally expected in the atmosphere. The decrease of the radio refractive index with altitude causes a downward curvature of horizontally directed radio waves. The radius of curvature of the rays is normally about four-thirds of that of the earth.[1] Under unusual meteorological conditions, however, the radius of curvature of the radio waves may be equal to or less than that of the earth (Figure 7-1), and the radio waves become trapped along the earth's surface. Under these conditions the energy is transmitted far beyond the normal radio horizon. Consequently, surface targets, prominent land features, or low-flying targets may be seen at distances of hundreds of miles with a radar which is reasonably sensitive. Moreover, the radar measurement of the altitude of these targets will be completely erroneous if the observer is unaware of the anomalous propagation situation. Several instances of UFO's as detected with radar have been identified later as being caused primarily by anomalous propagation.[2]

*Birds* have been identified or detected with radar from the mid-1940's. *Insects* are detected by sensitive radars but have been rather difficult to identify because of their small size and radar cross-section. Sin-

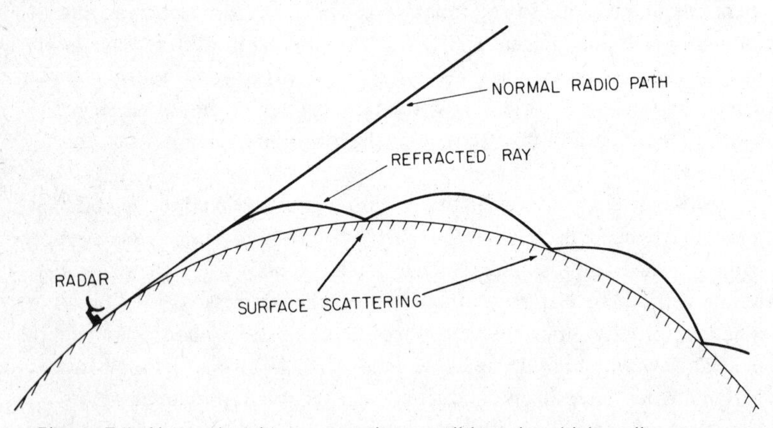


*Figure 7-1.* Unusual radar propagation conditions in which radio waves are multiply scattered along the earth's surface making many bounces, and which are propagated for much greater distances than the radar operator normally expects.

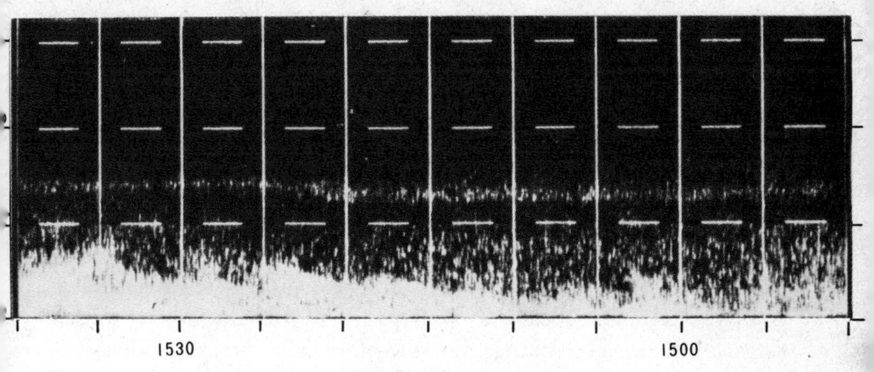


gure 7-2. Dot angels as seen with a vertically pointing 0.86 cm radar at Bedford, Mass., ember 1, 1966. The full vertical lines are at 5-minute intervals and the horizontal dashes espond to 5,000-foot height markers. Because of the way the TPQ 11 signal is recorded, a target appears as a short vertical line. The dot angels are caused by individual insects or s. On this occasion the concentration of insects near the surface is sufficiently large that arly solid radar echo occurs in the lowest 1,000 feet of the atmosphere. From Hardy and , and courtesy W. H. Paulsen and P. J. Petrocchi, Air Force Cambridge Research Labs., ord, Mass.

gle insects or birds may appear as dots on radar scopes, and the echoes have usually been called dot angels (Figure 7-2). The radar meteorological literature on dot angels of rather small cross-section was characterized by considerable confusion until about 1966. Some authors argued that the dot echoes were caused by insects and others argued that they were due to scattering from thin smooth surfaces across which an extremely large change in refractive index occurred. This confusion was largely eliminated, however, when multiwavelength ultrasensitive radars, capable of detecting a single large housefly at a range of 20 kilometers, were used to investigate the source of the dot angels. It was found that most, if not all, of the dot targets were due to single insects or birds. A description of these dot targets and the reasoning which led to their identification are given by Hardy and Katz.[3] As they pointed out, insects may occur at heights up to at least 4 kilometers. On occasion, the insects were sufficiently plentiful to serve as tracers of the air flow for an investigation of the low-level nocturnal jet stream, the passage of cold fronts, or convective processes in the clear atmosphere.

*Backscattering from fluctuations of the index of refraction* in the clear atmosphere is readily detected with ultrasenstiive radars having wavelengths in the decimeter range. These clear-air echoes generally occur in narrow layers which have considerable horizontal extent (Figure 7-3). The layers correspond in height to regions having sharp vertical gradients in refractive index, and the echo intensity can be accounted for by the refractive index characteristics which have been measured directly in the clear atmosphere. A recent investigation (by Hardy and Katz) of clear-air structures and processes was made with the powerful radars located at Wallops Island. Descriptions are given of convective thermals, hexagonal convective cells, breaking waves, and stratified layers in the clear atmosphere. It has also been determined that the clear-air radar layers in the 6- to 12-kilometer height range are characterized by turbulence sufficiently intense to affect aircraft.

The sources of clear air radar echoes are now well understood. Certainly there is no longer any question about the source or mechanisms which can give rise to the clear-air radar echoes, although the mystery and controversy surrounding the explanation of the "angel" echoes persisted from the early 1950's to about 1967. The problem was solved as soon as new and more powerful radars at several wavelengths were used in intensive investigations.

Strange and puzzling radar echoes have been detected in the past which, after considerable effort and study, have been explained. Assuredly, equally strange and bewildering radar echoes will be seen occasionally as new radars are put into operation or as the existing radars continue to carry out their remote probing mission. If a lesson has been learned from past analysis of mysterious radar echoes, it is that strange echoes or radar phenomena are rarely assessed, identified, or explained correctly when observed for the first time or for a short interval. Understanding of the mechanism responsible for the strange echoes comes only

---

*Figure 7-3.* Photographs of range-height indicators (RHI). The photographs were taken at 3.2-, 10.7-, and 71.5-cm wavelengths (top to bottom) along an azimuth of 260 degrees for 1740 EST, September 3, 1966, at Wallops Island, Va. The cirrus cloud appears at the shorter wavelengths, whereas the longer wavelength detects only the clear-air variations in refractive index. The numerous dot echoes which appear uniformly distributed between 1 and 3 kilometers at the two shorter wavelengths are due to single insects. From Hardy and Katz.

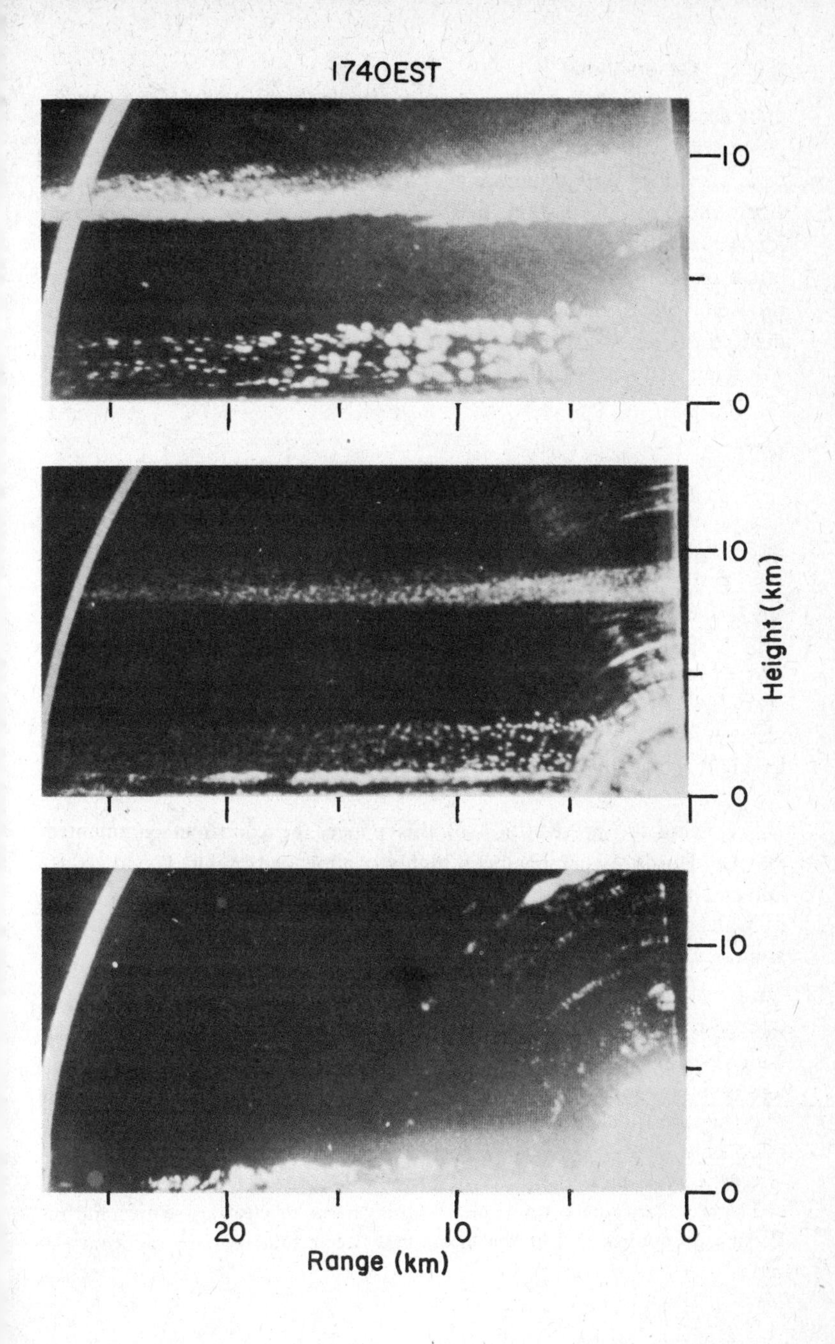
1740EST
10
0
10
0
Height (km)
10
0
20
10
0
Range (km)

after repeated observations and usually a painstaking analysis. The single observation of an unknown event with one radar is open to such a wide variety of interpretations that little is gained by proceeding on the single piece of information. It is necessary to take into account the various possibilities for the explanation of the strange echo, including the effect of the performance of the radar and recording system, and then proceed with a plan for a well-designed experimental (and possible theoretical) investigation of the event, and to hope that the phenomenon will recur in a manner which lends itself to study.

## NOTES

1. B. R. Bean and E. J. Dutton, *Radio Meteorology,* National Bureau of Standards Monograph 92 (Washington, D.C.: U.S. Government Printing Office, 1966).

2. E. U. Condon, *Scientific Study of Unidentified Flying Objects* (New York: Bantam Books, 1969).

3. K. R. Hardy and I. Katz, *Proceedings of the Institute of Electrical and Electronic Engineers* 57 (1969): 468–480.

## EDITORS' NOTE

Following the presentation of this paper, the chairman commented that Dr. Hardy's work has been highly commended by Dr. David Atlas, Director of the Laboratory for Atmospheric Probing at the University of Chicago, who sent his regrets that he could not attend this symposium. Dr. Atlas wrote:

> My own view is that while some of the UFO observations require almost incredible atmospheric structures for their explanation on the basis of propagation phenomena, some phenomena which were incredible just a few years ago have now been accepted by the community at large. Thus I fully expect that these still incredible atmospheric stuctures will be found to be entirely reasonable some years hence when our observational capacity can demonstrate their existence.
>
> In particular, I have reference to observations of clear-air scatter and reflection phenomena as reported in the last two to four years by radar meteo-

rologists and radio propagation scientists. Indeed, my own observations in the summer of 1969 at San Diego demonstrate the following: 1) the existence of exceedingly strong radar and radio-scatter layers with reflectivity ranging up to $10^5$ times as strong as ever reported previously, and with a thickness of only a few meters and often thinner; 2) the almost ubiquitous presence of wave motions on the layers in question. In view of the previously demonstrated facts that these layers both scatter and (specularly) reflect radio waves in the forward direction, there is now abundant evidence that the atmosphere will effect radar propagation in almost unbelievable ways and produce virtual targets which have apparently fantastic maneuverability.

In short, this is the sort of evidence which needs to be aired.

Some recent relevant papers are D. Atlas, F. I. Harris, and J. H. Richter, "Measurement of point target speeds with incoherent non-tracking radar: Insect speeds in atmospheric waves," *Journal of Geophysical Research, 75,* 7588 (1970); D. Atlas, J. I. Metcalf, J. H. Richter, and E. E. Gossard, "The birth of 'CAT' and microscale turbulence," *Journal of Atmospheric Science, 27,* 903 (1970); and E. E. Gossard, J. H. Richter and D. Atlas, "Internal waves in the atmosphere from high-resolution radar measurements," *Journal of Geophysical Research, 75,* 3523 (1970).

# 8

## Motion Pictures of UFO's

R. M. L. BAKER, JR.

The data that I have reviewed and analyzed since 1954 lead me to believe that there is substantial evidence to support the claim that an unexplained phenomenon—or phenomena—is present in the environs of the earth, but that it may not be "flying," may not always be "unidentified," and may not even take the form of substantive "objects." I would, therefore, prefer the label "Anomalistic Observational Phenomena" rather than "UFO." In this report, I will concentrate on the anomalistic observational phenomena as depicted in motion pictures, and will not attempt to support any particular hypothesis as to the source of the phenomena. I will show and analyze four film clips, and discuss two others in a brief fashion. Two of these films—the Montana 1950 and the Utah 1952 films—have been dealt with rather thoroughly in the past. The third was taken by Policeman William Fisher on March 9, 1967, in Moline, Illinois, and has not, to my knowledge, been as thoroughly analyzed as the first two clips. The fourth film was taken by Clifford C. DeLacy at Kaimuki, Honolulu, Hawaii, on January 3, 1958, and I do not know of any thorough analysis.

I believe that these film clips are rather typical of the anomalistic or UFO motion pictures. Although I am convinced that many of the films indeed demonstrated anomalistic phenomena, they all have the characteristic of rather ill-defined blobs of light, and one can actually gain from them little insight into the real character of the phenomena. For example, linear distance, speed, and acceleration cannot be determined precisely, nor can size and mass. This situation is not particularly surprising, since, without a special-purpose sensor system expressly designed to obtain information pertinent to anomalistic observational phenomena, or a general-purpose sensor system operated so as not to disregard such data, the chance for obtaining high-quality hard data is quite small.

The films are rather ungratifying subjects for research, because of their low information content (they simply show little dots of light) and because their analysis must often rely, in part, on the soft data of eyewitness reports. Only two generalizations can be made: the photographic images usually occur in pairs and usually exhibit a slightly elliptical form.

## Montana 1950 Film

Two anomalistic unidentified flying objects were sighted and a few moments later photographed at about 11:30 A.M. MST in August 1950 (exact date is uncertain, but shadows on the film confirm the time of day given by the witnesses) by Nicholas Mariana at Great Falls, Montana. Mr. Mariana owned and operated a radio station in Missoula, Montana, and was owner of the Great Falls baseball team. He habitually kept a movie camera, used in the UFO photography, in the trunk of his car. All of the soft data (eyewitness reports of Mr. Mariana and his secretary) indicated the objects were silvery in appearance with a notch or band at one point on their peripheries and could be seen to rotate in unison, hover, and then "with a swishing sound, floated away to the left [*South*]." The film itself is disappointing and only shows two inarticulate bright white dots, which passed behind a water tower.

Figure 8-1 is adapted from a land survey of the area made by Henen Engineering Company, and indicates the location of the Anaconda smokestack. Mariana and his secretary were looking toward this smokestack while standing ten to fifteen feet in front of the turnstile on the right of the figure when they first viewed the objects moving from the north toward the water tower. (He had been looking at the smokestack in order to determine the wind direction.) The movie-camera directions of the first frame and the last frame are shown in Figure 8-1,

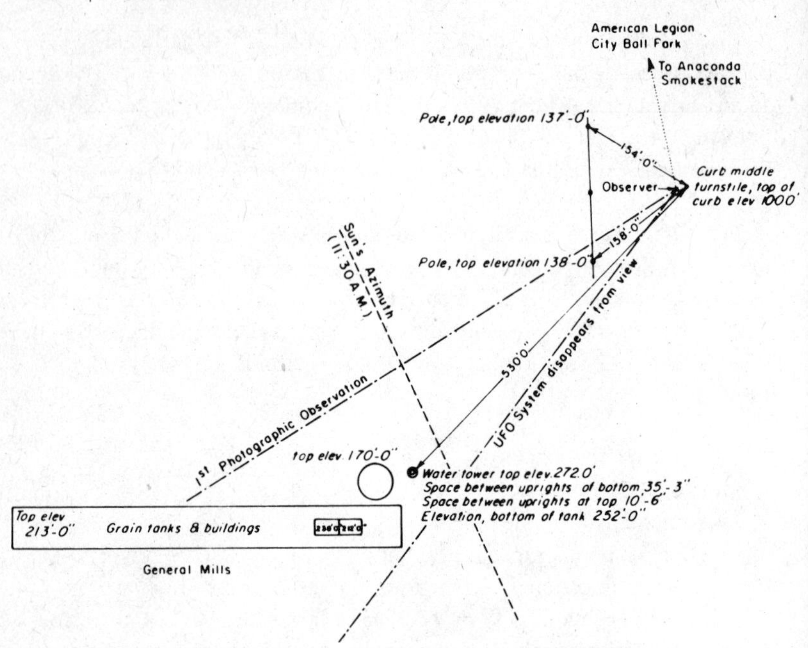


*Figure 8-1.* Land survey of the filming region of the Great Falls, Montana, sighting. Adapted from a survey by the State of Montana Land Surveyors.

as determined by iconolog (a film viewer with movable cross-hairs and a digitalized output) measurements.

The path of the objects as they passed behind the water tower is shown in Figures 8-2 and 8-8. The angular data were obtained as noted by utilizing the reference points marked 3, 5, and 6.

Figure 8-3 shows the manner in which the diameter of the bright objects decreased with time. These measurements made by the author are the least accurate of the data because of the smallness of the dimensions and the fuzziness of the images. The image of any light source as seen by either the eye or a camera can appear much larger than the source itself. This fact had obvious bearing on the analysis of the film, so I undertook a photographic experiment during December 1955. The experiment was designed to obtain empirical information on the effects of distance, lens focal length, iris stop, frame speed, and other factors on the

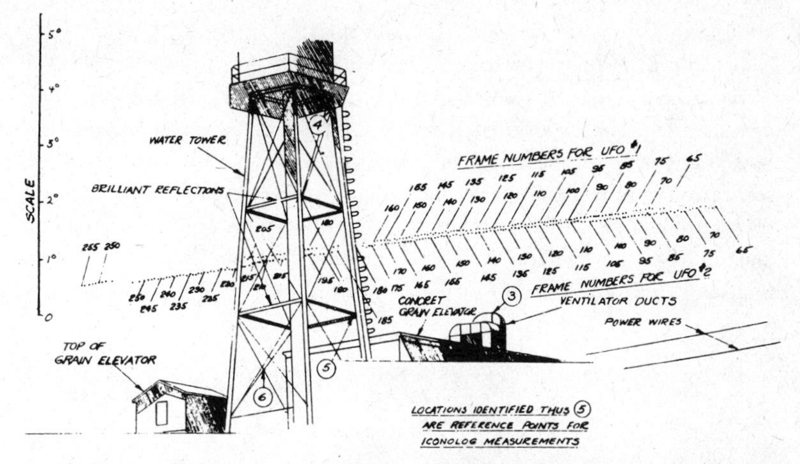


*Figure 8-2.* Trajectory of anomalistic unidentified flying object observed in the Great Falls, Montana, motion picture film with frame numbers indicated. Drawn by the author.

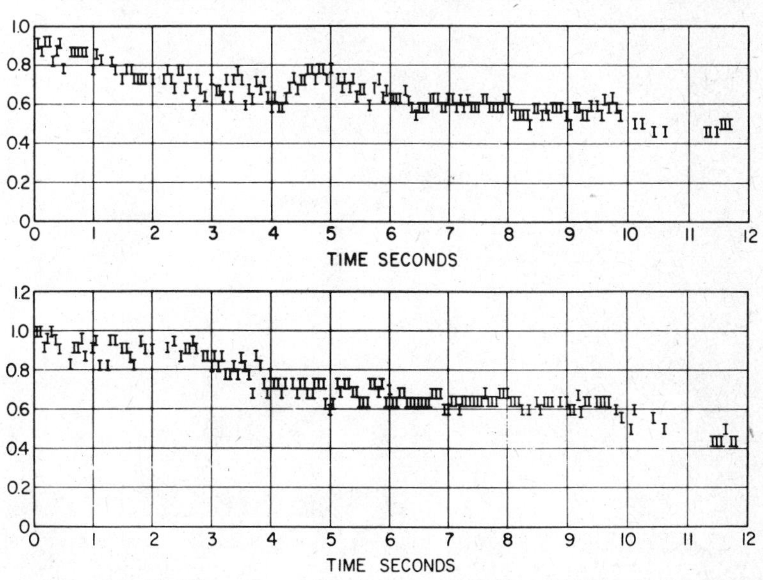


*Figure 8-3.* Variation of angular diameter of UFO's with time in the Great Falls, Montana, motion picture. Drawn by the author.

photographic images of various bright sources of sunlight. Some 118 combinations of these variables were examined.

Figure 8-4 shows the variety of diffuse and specularly reflecting objects chosen. The experimental results, using a camera similar to Mariana's, appeared to indicate that if the first few frames of the film show sun reflections from, say, airplanes which were optimally oriented with respect to the sun, then the planes would have been one to three miles distant from the camera. If, however, these first few frames represent images of the reflections of airplanes *not* quite optimally oriented, then

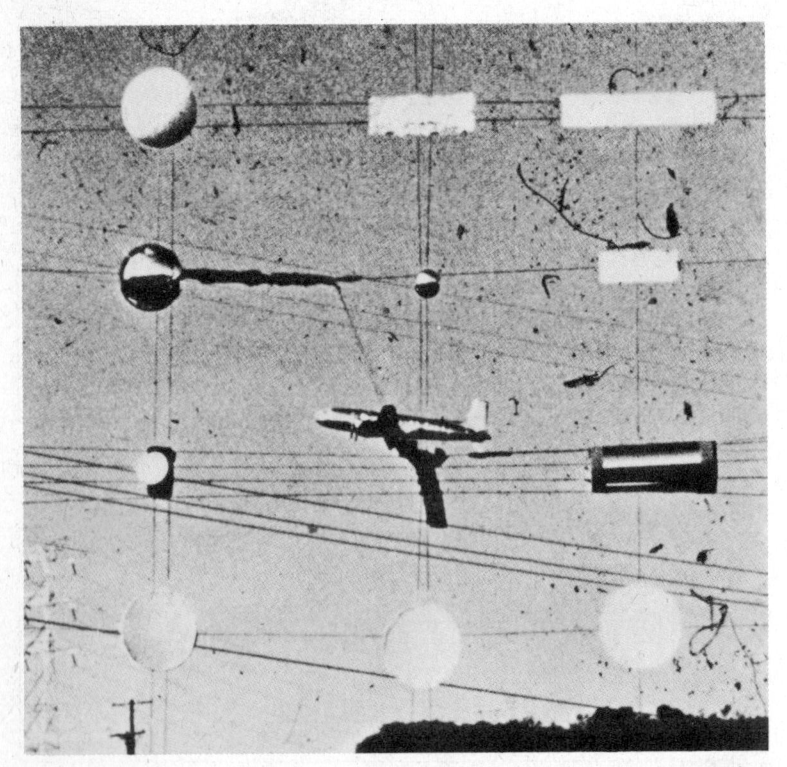

*Figure 8-4.* Control reflectivity experiments with motion picture camera similar to that used in the Great Falls, Montana, sighting. Obtained by the author as described in the accompanying text.

the planes could have been closer. In either event, their structure would have been visible.

Figure 8-5 compares an enlargement of one of the frames of the Montana film (a copy of a copy, at reduced contrast) with a frame taken during the photographic experiment showing jet planes, optimally oriented with respect to the sun. The light reflection images are comparable to those found on the Montana film, but the structure of the aircraft is clearly visible.

*Figure 8-5.* An enlargement of one of the Great Falls frames (left) compared with an experimental frame of jet aircraft obtained by the author. Both photographs are to the same scale.

The Montana film contains six independent quantities that vary with time through about 225 frames (frames 65 to 290). These quantities describe the UFO images: the two degrees of freedom of each dot measured from the film after the foreground appears, on frame 65, and the two apparent diameters of the developed image as measured on all 290 frames. In the analysis it was convenient to treat the two UFO's as a system. The four degrees of freedom chosen for this system were the azimuth and altitude of the midpoint on the line of centers between the UFO images, their angular separation, and their inclination to the horizon. The inclination angle was found to be very small, the objects appearing to move almost in a plane parallel to the ground. There is a slight decrease in the angle of inclination as the objects regress, but its small value is almost masked by random errors inherent in the measurements.

Figure 8-6 is a time-plot of the angular altitude, *h,* and the azimuth,

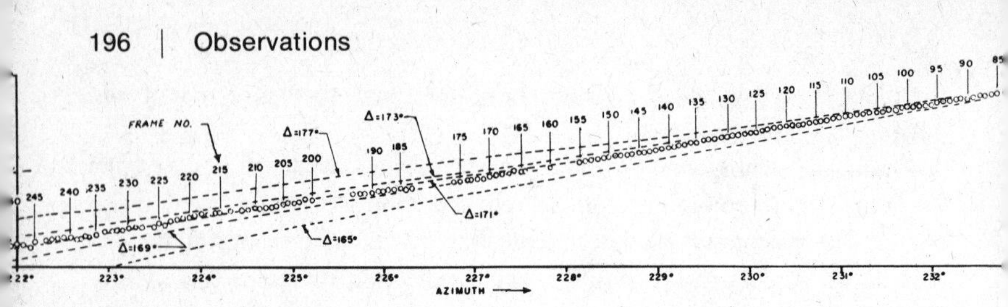


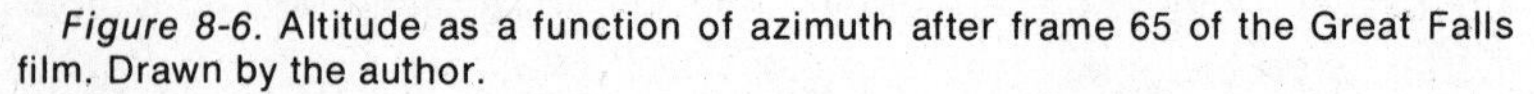

*Figure 8-6.* Altitude as a function of azimuth after frame 65 of the Great Falls film. Drawn by the author.

$A$, of the midpoint of the line of centers after frame 65, and Figure 8-7 shows the separation-distance ratio $\theta_0/\theta$ as a function of time, where $\theta_0$ is the initial angular separation on frame 1 and $\theta$ is the angular separation at any given time. Some frames were not measured, due to the obscuration of the images behind the water tower. There were two frames missing between frame numbers 177 and 180 on the 35 mm print that was measured for separation distance, but these were accounted for in the time scale, using the 16 mm original as a basis. About 225 frames after the foreground (ventilator duct) appears on the film, the objects can no longer be clearly identified, and measurements become very uncertain.

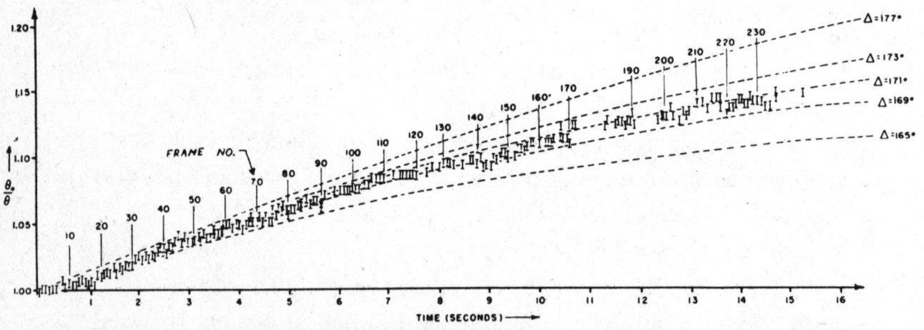


*Figure 8-7.* The angular separation distance of the system of UFO's as a function of time for the Great Falls motion picture. Drawn by the author.

In Figures 8-6 and 8-7 the dotted lines represent what would be the locus of data points if the objects remained the same linear distance apart, and moved linearly in a horizontal plane on headings, $\Delta$, of 169° to 177°. All of the data seem to be consistent with the foregoing assumptions and a heading of 171°. Of course, one cannot absolutely rule out some other curvilinear motion of the objects. However, any such motion would necessitate the coincidence of azimuth, altitude, and separation, all varying proportionally in some very peculiar fashion to a tolerance of one per cent.

Figure 8-8 is a map of Great Falls, Montana, showing the motion of the UFO's at various hypothetical distances. (No absolute determination of distance can be made on the basis of the angular data.) It also shows where Mariana and his secretary first viewed the "hovering and rotating" near an Anaconda smokestack.

W. K. Hartmann also analyzed the film clip and investigated the ellipticity of the images (which I had originally attributed to blurring caused by irregular panning). He computed the apparent inclination, $i$, of the disk-shaped images, where $i = \sin^{-1}(b/a)$ with $a$ and $b$ being the major and minor axes of the elliptical image. Hartmann found (see Table 3, p. 414, of the Condon Report) clear evidence of camera motion in frame 2, but that otherwise there was a constant ellipticity or

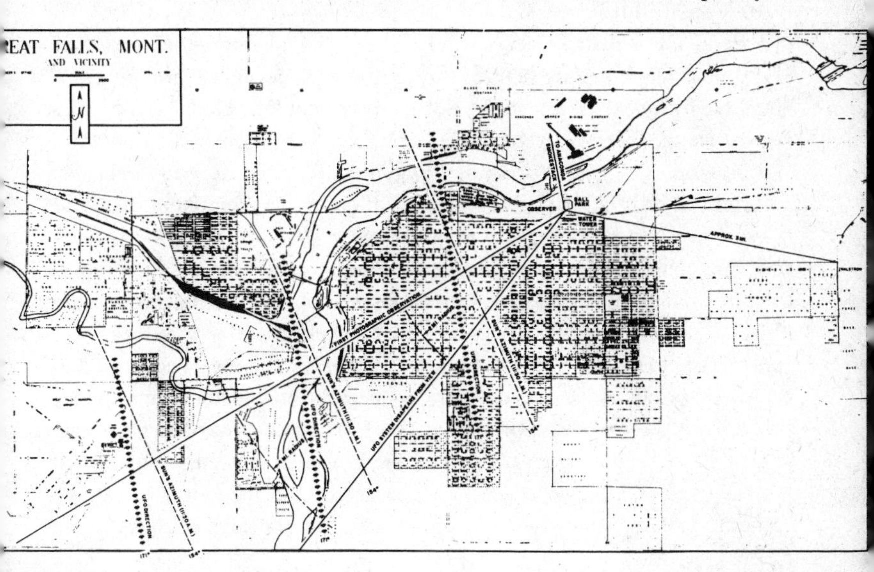


*igure 8-8.* A map of Great Falls, Montana, indicating possible UFO motions for a variety of sumed distances. Dotted lines are assumed tracks; dashed lines are sun's azimuth; solid es are viewing lines.

flattening of the fuzzy images, with $a = 1.15b$, or $i = 60°$. Such ellipticity would be consistent with an oblate form such as a disk, or with a consistently flared reflection. The data are not really precise enough to make a firm hypothesis.

After more than a decade of speculation and hypothesis checks, all natural phenomena (birds, balloons, insects, meteors, mirages, and so forth) have been ruled out. (Since the date of the photograph is uncertain, weather bureau reports are not pertinent, but the uniform motion does not seem to be consistent with balloons.) The main possibility is that of airplane reflections or, perhaps, some airplane-related phenomenon such as luminous shock waves. The airplane hypothesis may seem attractive, but it does not really jibe with my analysis or with Hartmann's. In short, planes at the largest distances compatible with their speeds and the angular rate of the images would have been identifiable on the film.

### Utah 1952 Film

Several anomalistic objects were sighted and photographed at about 11:10 A.M. MST on July 2, 1952, by Delbert C. Newhouse at a point on State Highway 30, seven miles north of Tremonton, Utah (see p. 13). Newhouse, a Chief Warrant Officer in the U.S. Navy, was driving from Washington, D.C., to Portland, Oregon, with his wife and their two children. Shortly after passing through the city of Tremonton, his wife noticed a group of strange shining objects off toward the eastern horizon. She called them to her husband's attention and prevailed upon him to stop the car. When he got out he observed twelve to fourteen of the objects and was sufficiently impressed by their peculiar appearance to run to the trunk, get out his camera, and begin filming. There was no reference point above the horizon, so he was unable to estimate size, speed, or distance. He reports that one of the objects reversed its course and proceeded away from the rest of the group; he held the camera still and allowed this single object to pass across the field of view of the camera, picking it up later in its course. He repeated this for three passes.

The images on the color film are small and relatively sharp, as confirmed by Hartmann's analysis (p. 422 of the Condon Report), but with no background, they are difficult to identify. Figure 8-9 shows two plots of relative motions, each utilizing one of the objects as a reference point. There is a general tendency for the objects to move in pairs—although not nearly as uniformly as the Montana objects. Figure 8-10 is a blowup of one of the frames of the Utah film and exhibits the pair configuration.

The tendency for a camera man to pan with the objects would yield an underestimate of their angular rate. Our measurements show that if the objects were at, say, 2,000 feet they would be moving about 9 miles per hour and pulling about 0.25 *g* acceleration relative to one another.

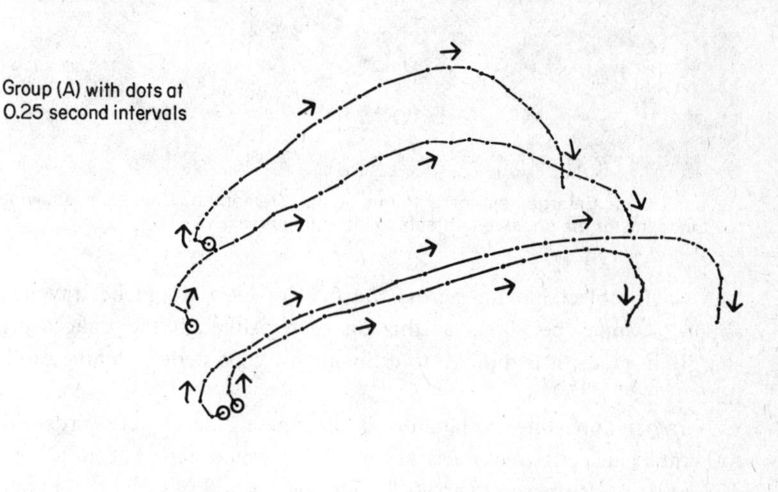


*Figure 8-9.* Plots of relative motions of unidentified flying objects in the Tremonton, Utah, film. Drawn by the author.

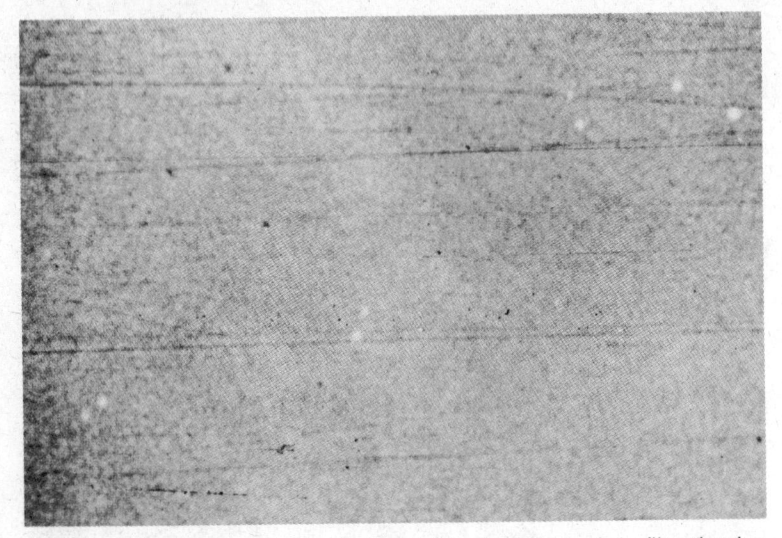

*Figure 8-10.* Enlargement of a frame of the Tremonton, Utah, film showing the tendency for the observed objects to move in pairs.

The single object moving across the field of view would be traveling about 54 miles per hour at this distance. Although the images are sharply in focus, it is difficult to estimate any consistent flattening or ellipticity.

A rather appealing explanation is that these objects were birds. On the other hand, this motion is not what one would expect from a flock of soaring birds; there are erratic brightness fluctuations, but there is no indication of periodic decreases in brightness due to turning with the wind or flapping. No cumulus clouds are shown on the film that might betray the presence of a thermal updraft. In addition, there is the soft-data question of why a person would be so struck by a flock of birds milling about that he would go to the trouble of photographing them. Hartmann says that he has witnessed flights of white gulls at Tremonton in motions that might have duplicated those on the Utah film, but that he had no camera to verify this. I have never seen bird formations so striking that I would not recognize them as birds, or so unusual that I would film them. The motion pictures I have taken of birds at various distances have no similarity to the Utah film. Thus, to my mind, the bird hypothesis is not very satisfying and I classify the objects as anomalistic observational phenomena.

## Illinois 1967 Film

The Illinois film was taken by William Fisher, a patrolman with the Moline Police Department, at about 1:30 P.M. on March 9, 1967, near the intersection of 14th Street and 16th Avenue, Moline, Illinois. In a telephone interview, Mr. Fisher told me that when he first observed the object just above and to the north of some trees, moving west, he thought that it was an airplane coming toward him. He then noticed that it was just hovering, had a "football" shape, and exhibited a "gold-like" reflection, and a "machined or tooled" surface with a definite shadow underneath. The sun was near the zenith. Fisher dismounted from his motorcycle (leaving the motor idling) in a state of disbelief. He remembered that he had his camera available, a Holiday 8 mm with a Mansfield turret (which leaves uncertain which lens was employed), and began filming. As he continued filming, he said that the object "drifted" away to the northwest and finally became "infinitely small" and disappeared from view after a few minutes (and after Fisher had run out of film). Fisher then said he viewed a second object about one or two minutes later. It followed about the same path as the first, and he attempted to film it, but his film was exhausted.

Fisher said that a number of other people had observed both of the anomalous images. He asked them what they thought they had seen in order not to "prejudice their testimony." A Mrs. Leo Schmitz indicated to Fisher after the first object had passed that she had seen an oval-shaped object moving in the sky and had heard a sound "like air escaping from a tire." (Fisher heard nothing except the idling noise of his motorcycle, leaving as the most plausible explanation that there was indeed air escaping from a tire.) A Mrs. Frank Daebellihn, several nuns, and a number of children indicated that they had seen two strange objects.

I discussed the sighting with Mrs. Schmitz, who confirmed that she had seen the object but said that she noticed no metallic or textured surface. It appeared to her to be white, and did not shine as if by reflection or self-illumination. (Fisher agreed that "On my initial viewing the object did appear to be off-white or dull silver and did not reflect as intensely as it did during later viewings.") Mrs. Schmitz said that it appeared to be more elliptical than a football, "at least twice as wide as it was high." She was sure that it hovered and then moved off. She stated that she could see it "very clearly, more clearly than the movie . . . showed it." She was not positive that it was not a bird or an airplane, but she could not concede that it could have been a balloon or a blimp. The sighting "absolutely amazed" her. She stopped her car near Fisher and felt compelled to watch the object, then left after the first object passed and was not present to confirm the presence of the second. She wrote to Allen R. Utke of Wisconsin State University about the sighting, and Utke also received a letter from Mrs. Frank Daebellihn, who wrote that the object was "oblong" and "real shiny," but she also admitted to "not too good eye sight."

The UFO film clip is, typically, very disappointing and only shows a small oval "blob" of light decreasing in size against a plain blue-sky background. There are little hard data present on the film. Microscopic examinations show a definite ellipticity on the first few frames, but it is masked by poor atmospheric seeing. Figure 8-11 is a copy of a frame supplied by Professor Utke. The elliptical inclination, *i,* varies between about 18° and 45° on the initial frames. Fisher could only estimate it at about 45°. He was adamant that the objects were not balloons because they appeared to be heading into a 30- to 40-mile-per-hour wind, and he seemed equally certain that they were not birds or airplanes.

Professor Utke and William Powers of Northwestern University had studied this sighting, and I discussed it with them. Utke was very familiar with the area and found that there had been a number of related visual sightings there at about the same time. (Several of these were documented by R. B. Dyke, Director of a UFO Research Committee, which he formed.) Utke had contacted the Midwest Central Weather Bureau and the Moline Airport; there was no weather balloon launching near the time of the filming, and the winds were 15 knots SSW at the surface, 30 to 35 knots at 3,000 feet, and 25 to 30 knots at 5,000 feet WSE. The object's reported motion was east to west, more than 90° off the wind direction. Utke suspected that the anomalistic phenomena

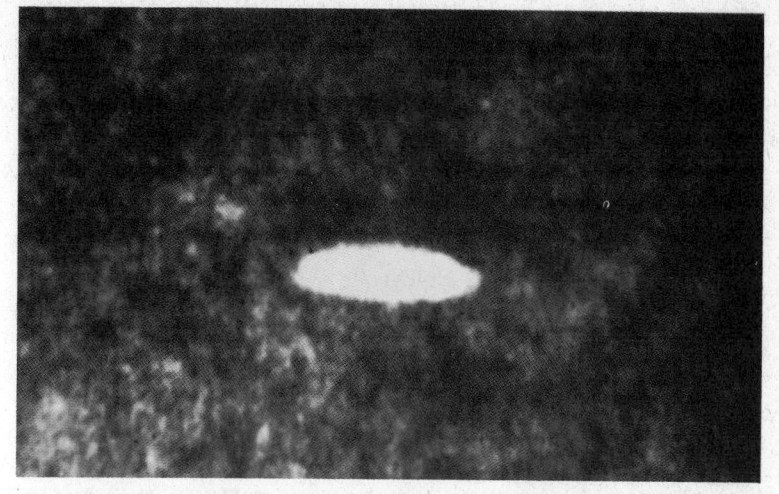

*Figure 8-11.* Enlargement of a frame of the Moline, Illinois, film. Photograph provided through the courtesy of Professor Allen Utke.

"had something to do with the Rock Island Arsenal," primarily because of the arsenal's proximity.

Powers had viewed the film several times, without detailed study, and he was convinced that the object was not Venus and not an airplane. He had learned at the Moline Airport that there were C-131 aircraft and heliocopters there, but that none was airborne at or near the time of the sightings.

About all that can be concluded from the Illinois film is that an elliptical image was photographed, that its angular size gradually decreased, and that there was no periodic or sudden change in its luminosity. Because of the length of the filmed sequence and the uniformly changing "brightness" of the object, the airplane and bird hypotheses are difficult to support. Balloons cannot really be ruled out on the basis of the elliptical image, although the witness reports them unlikely.

## Hawaii 1958 Film

The Hawaii film was taken by Clifford DeLacy, at that time a student majoring in nuclear physics at Vallejo Junior College. He related that the film was taken at about 4 P.M. on January 3, 1958, in his mother-in-law's back yard near Harding Avenue and 6th Street, Kimuki, Honolulu, Hawaii, where he was on vacation. He was relaxing in the back yard when he was startled at the appearance of some nine objects "flashing across the sky in a northwesterly direction grouped in pairs." They were estimated to be about 40° above the horizon. He called to his wife and then went into the house to fetch his camera equipment, a tripod and an 8 mm Revere movie camera with a 1½" telephoto lens. Exicitedly he began filming without the tripod. After a few seconds he calmed down, set the camera on the tripod, and completed the filming. At the beginning, DeLacy was successful in shooting the tops of trees in order to establish a reference point for the angular rate and, after a survey, the angular altitude and azimuth.

DeLacy had made a hobby of studying UFO's and stated that "in my opinion, a good 90 per cent of all the reports concerning 'flying saucers' are mere hallucinations—or worse." His color film involves two very brief sequences showing anomalous images sandwiched in between conventional amateur travel film sequences. It starts with a typical Hawaiian sunset scene and closeups of flowers. The dull and degraded nature of the color and the lack of contrast indicate typical poor quality of a copy of amateur movie color film. After the flowers, there is a blue sky with traces of cumulo-nimbus cloud formations and some tree tops (DeLacy estimates that he was about 15 to 30 feet from the trees and that they were about 15 feet high). After about 10 seconds, two very fuzzy moving objects can be barely identified moving from right to left. Apparently, DeLacy then placed his camera on the tripod, and one can see a wiggly line on the right for just a few frames. There is no evidence on the film that an object made a sharp turn; it might have, but it is more reasonable to suppose unsteady handling of the camera. At this point all foreground disappears, and one very bright image can be seen moving across the sky and a glimpse of a second object. The camera is still a bit jittery, but microscopic examination tends to show that the elliptical image is not entirely due to irregular panning. The whole UFO sequence lasts but 25 or 30 seconds, although DeLacy says that the entire visual sighting lasted about 5 minutes. It is followed by other home-movie scenes. As is usual, the film is *not* a dramatic and exciting portrayal of anomalistic phenomena by any means. Actually the information content is a bit better than one would expect, primarily because of the foreground. The angular velocities appear rather high, partly because of the 1½″ telephoto lens (3″ lenses were used for the Montana and Utah films). By running the film backward and taking great care to identify the very fuzzy initial images, one can establish an angular velocity with respect to the trees that is between 0.03 and 0.12 radians per second (larger than the Utah film's 0.01 to 0.07 radians per second, relative, and the Montana film's 0.02 radians per second, absolute). A more careful study could establish the angular rate rather precisely (each frame covers about 0.125 radians in breadth, and the images appear to move in some frames at one frame-width per 16 frames with respect to the foreground features). More analysis is needed of the assumed constant frame speed (16 frames per second), and a survey of the tree locations would have to be made in order to substantiate this tentative conclusion. If the high angular rate is correct, then the UFO's would have been moving transversely at about 900 miles per hour at a distance of 2 miles. At this distance aircraft would have been identifiable, as I discussed in the Montana film clip analysis. The clarity of the tree tops probably indicates a good focus. At 2,000-feet range, birds would have to have been moving at about 170 miles per hour transversely.

DeLacy reported that the objects were solid, dull to bright. The only other witnesses were his wife Jacqueline and his mother-in-law Mrs. Abel M. Rodriguez. The marine airfield tower operator reportedly told DeLacy that no jets were in the area. DeLacy feels strongly that he was not photographing birds. He estimated that some of the anomalous objects moved three to five times faster than a jet might move at a distance of a mile or two.

## Other Films

About two-thirds of the UFO films (or films that purport to show anomalistic observational phenomena) that I have viewed have been hoaxes or obviously conventional phenomena, or after analysis, an unusual film of a natural phenomenon. I have chosen in the foregoing discussion some examples of films that I interpret as involving anomalous observational phenomena. There exists other photography I would like to mention next but, unfortunately, very limited analysis of this photography has been accomplished to date.

### Florida 1955 Film

During the course of the analysis of the Utah and Montana film at Douglas Aircraft Company, I had the opportunity to view gun-camera photographs taken over Florida. Unfortunately, we could not retain this film, and did not have time to do an analysis, and determine the exact circumstances of the filming, e.g., exactly when and where it was taken and the details of the gun camera and the pilot's reactions.

The Florida film was disappointing; it showed only a pair of white-dot images. However, since a foreground was present, an accurate study could have been carried out. The director of the Douglas Aircraft research office, Dr. W. B. Klemperer, agreed with me on a preliminary conclusion—not supported by detailed analysis—that no ordinary natural phenomenon was a likely source for the images.

### Venezuela 1963 Film

In June of 1963 I received a movie film clip from Richard Hall of NICAP that had purportedly been taken from a DC-3 aircraft near Angel Falls, Venezuela, at about 12:15 P.M. This clip was 8 mm color film, exposed at 16 frames per second, and showed a very bright yellow, slightly pear-shaped object that disappeared in a cloud bank after 60 or 70 frames. At the time I was head of the Lockheed Aircraft Company's Astrodynamics Research Center, where two photogrammetricists, P. M. Merifield and James Rammelkamp, were able to undertake a study of the film. They found little of interest, and after their preliminary examination I expended considerable effort in further analysis. Again, I was only able to conclude that the yellow object was no known natural phenomenon; but before we could make accurate measurements of angular rates and acceleration, the film was lost. We had one microphotograph of the object on one frame shown in Figure 8-12.

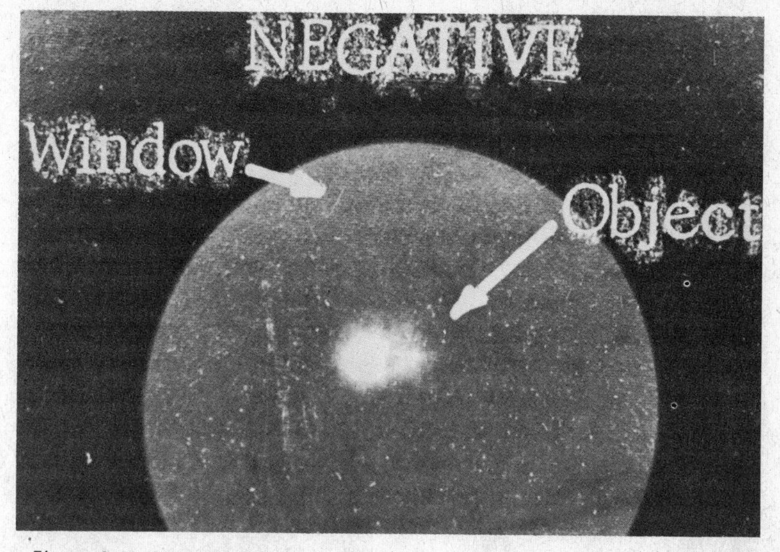


*Figure 8-12.* Microphotograph of the 1963 Venezuela object—the one frame remaining after the loss of the film.

## Oregon 1950 Pictures

This filming took place on a farm just south of U.S. Highway 99W and southwest of McMinnville, Oregon, on May 11, 1950.* One witness, who reportedly saw a metallic-looking, disk-shaped UFO, called another from the farmhouse and after excitedly searching for their camera the second witness took two still shots (Figure 8-13), both of which included overhead wires, although these are not visible in the reproduction. Hartmann carried out a thorough analysis (see pp. 396–407 of the Condon Report). His conclusions are as follows:

> This is one of the few UFO reports in which all factors investigated, geometric, psychological, and physical appear to be consistent with the assertion that an extraordinary flying object, silvery, metallic, disk-shaped, tens of meters in diameter, and evidently artificial, flew within sight of two wit-

* Since Hartmann's comprehensive analyses for the Condon Report, there have been some questions raised with respect to the time of the photography.

*Figure 8-13.* Two views of the McMinnville, Oregon, film's UFO. Courtesy of United P International.

nesses. It cannot be said that the evidence positively rules out a fabrication, although there are some physical factors such as the accuracy of certain photometric measures of the original negatives which argue against a fabrication.

## England 1971 Film

At about noon on December 15, 1971, six members of an ATV film unit observed an object, which according to the cameraman, Noel Smart, was "bright silver . . . changed to a bright fluorescent or luminous type of orange . . . hovering . . . and then flying of with a . . . vapor trail . . ." They recorded the object on 16 mm color film near Redford Bridge, Enstone, Oxfordshire, England. Typically, the film is difficult to analyze. One brief filmed sequence exhibits a broad trail, which I have never witnessed before. The object could be an aircraft dumping fuel or a conventional aircraft accompanied by a vapor trail.

Because of the color and motion of the filmed object much more analysis would be required in order to be more certain of these natural phenomenon explanations.

## Conclusions

As already mentioned, "UFO" films are ungratifying for research—at least those I have seen to date. Amateur photographic equipment is usually brought into action after the most remarkable aspect of the phenomenon has passed, the photographer is usually excited, his camera is not at hand, and he is ill prepared to do an adequate photographic job. Furthermore, films taken with amateur or even professional photographic equipment cannot be expected to be adequate for photogrammetric analysis as would, for example, cinetheodalite films. Thus, we find ourselves viewing images of little blobs or dots of light. About the only correlation among them is that the images are usually elliptical and usually come in pairs. The characteristics of these blobs and dots may rule out most natural interpretations, but they cannot define what really is being portrayed on the film. It is frustrating to analyze these films. One often wishes to grasp at some candidate natural phenomenon, only to find this first theory shaken and, in all honesty, to discover that the natural-phenomena hypothesis is faulty and should not be further maintained.

If the only alternatives to birds, airplane reflections, mirages, balloons, Venus, and so forth were little green men from another solar system scooting around in flying saucers, then one would be forced to say that such creatures and machines are so unlikely that *any* alternative, no matter how hard it is to justify, is "better." I do not hold to this concept of *one* alternative hypothesis. I believe that photos *are* hard observational data (albeit extremely vague in meaning due to low information content), data that result from some poorly understood phenomenon or phenomena. It may be that these photographed phenomena are related to ball lightning, or the rocket effect in small comets entering our atmosphere, or ephemeral natural meteoritic satellites of the earth, or a thousand other things. Whatever they are, we are obliged to find out more about them. It is my conclusion that there are only so many quantitative data that can be squeezed out of the vast amounts collected to date, including the "bit buckets" of surveillance-radar uncorrelated targets (UCT's).

I believe that we will frustrate ourselves by endless arguments over past, incomplete data-scenarios; what we need is more sophisticated analysis of fresh observational anomalistic data. We must come up with more than just a rehash of old data such as the fuzzy white dots shown in Figures 8-5A and 8-10.

It seems very unlikely that existing optical and radar monitoring systems would collect the type of quantitative data required to identify the phenomena. Moreover, we currently have no satisfactory basis upon which to evaluate the credibility of the myriads of eyewitness reports. Thus, continuing to "massage" past reports of anomalistic events would seem to be a waste of our scientific resources.

In balance, then, I conclude that we are not now, nor have we been in the past, able to achieve even partially complete surveillance of space in the vicinity of the earth sufficient to provide statistical information on anomalistic phenomenon. Hard data on anomalistic observational phenomena do exist, but they are of poor quality because of the equipment employed in obtaining them. Soft data on anomalistic phenomena also exist, most of them of doubtful credibility. Experiments should be devised, and study programs should be initiated, expressly to define anomalistic data better. In order to justify such experiments and associated studies, it is not necessary to presuppose the existence of intelligent extraterrestrial life operating in the environs of the earth, or to speculate about "their" advanced engineering capabilities or "their" psychological motivations.

# PART III

## SOCIAL AND PSYCHOLOGICAL ASPECTS

# 9

## Sociological Perspectives on UFO Reports

ROBERT L. HALL

I should like to turn attention away from sundogs and spaceships and focus it on human behavior. Reasonable men may disagree as to whether UFO reports imply the existence of an important, unfamiliar physical phenomenon worthy of study. However, we can agree that for many years, in all parts of the world, many people (including intelligent, reliable witnesses) have been reporting flying objects which they found puzzling. We can agree that their reports contain many recurrent features which, if taken at face value as reliable testimony, would suggest something other than conventional aircraft and meteorological and astronomical phenomena. We can agree that a great many people are interested and have become involved in trying to account for these reports, or perhaps more often, in defending a position about what might account for them. I am afraid that we must even agree that scientists have sometimes been caught up in the controversy and defended a position with more emotion than logic. We might differ as to which scientists are being emotional and which are being logical, but we are clearly witnessing some kind of phenomenon which stirs both scientific controversy and human emotion. There are clearly some important *behavioral* phenomena, though we may disagree about what *physical* events must be posited to account for the behavior we observe.

As a behavioral scientist I focus my attention on questions about these behavioral phenomena—reports of flying objects, elaborate beliefs about the reported objects, human controversy in defense of beliefs, and even the behavior of scientists in analyzing the reports. I begin with the plausible assumption that the thousands of reports do not have a single cause: they contain a potpourri of deceptions, delusions, and illusions, and at least some accurate testimony. Our basic problem is to sort out those components.

I shall organize my comments around three main issues. First, I shall summarize some knowledge about such processes as rumor and systems of belief, mass hysteria, and hysterical contagion as these may apply to reports of flying objects. Second, I shall discuss the plausibility of systematic misperception to account for the "hard core" of UFO sightings and the question of whether there exists a subset of UFO reports which requires other interpretation. Third, I shall take the risky step of commenting to the group of scientists present here on the behavior of scientists in response to UFO reports and other similar phenomena.

## Systems of Belief and Contagion of Belief

Nearly all rational observers agree that the great majority of reports of flying objects have their origin in misidentifications of familiar phenomena, together with a few hoaxes and delusions. The sky, especially the night sky, is full of ambiguous stimuli, and people generally show a powerful need to reduce ambiguity. Much research in sociology and social psychology indicates that the typical first reaction to an ambiguous event is an effort to explain it in terms of something familiar. This kind of improvised clarification is the essence of rumor. Thus, for example, following the explosion of the first atomic bomb over Hiroshima, the early rumors were: (1) the city had been sprayed with gasoline and set afire, or (2) a huge cluster of incendiary bombs had been dropped, or (3) a fine magnesium powder had been sprayed on the city and ignited by electric power lines.[1] The reaction is characteristic: given an ambiguous event and a lack of trusted information to explain it, people improvise explanations, trying first those explanations that require no really new knowledge but build directly on what they already believe.

J. Allen Hynek, in the process of interviewing hundreds of UFO witnesses, has observed a phenomenon which he has labeled the "escalation of hypotheses," which appears to be a specific instance of the general tendency to explain first in familiar terms. That is, in numerous cases the persons reporting a UFO have indicated that they first tried to fit their observations into familiar categories and came to regard the phenomenon as strange and unidentified only after its appearance and actions seemed clearly to rule out familiar interpretations, such as airplanes, helicopters, clouds, birds, stars, and planets. This is an important point and seems quite contrary to statements sometimes made by noted UFO skeptics, who refer to witnesses as eager to find something strange.

What people believe is usually organized into elaborate systems of belief. That is, each person has a cognitive structure consisting of many items of information and belief which are interdependent, and people are organized into social systems in which each person lends support to beliefs of others in the system. A lonely belief is an unstable belief; just as nature abhors a vacuum, nature abhors an isolated belief.

There has been very extensive research on cognitive structure and cognitive processes in recent years,[2] and any brief summary statements can be risky. However, it appears that people tend in most circumstances to hold beliefs consistent with those of people around them, because they interact selectively or are influenced or both. It appears that people perceive more readily and accurately those things that are consistent with their pre-existing knowledge and beliefs than things that are not. It appears that ambiguous situations tend to be interpreted so as to fit in with and support pre-existing belief and knowledge, and gaps in knowledge tend to be filled with consistent improvisations. When there is a strong system of belief with substantial social support, it is likely to be defended vigorously, beyond the dictates of logic. Conversely, when reasonable men report events which receive no social support from their friends and do not fit their own prior beliefs, we have to take these reports seriously.

Combining knowledge of reaction to ambiguity with knowledge of systems of belief, we expect that an ambiguous event will tend to become assimilated into a pre-existing system of belief. Suppose we take a case of someone committed to a system of belief which asserts man's basic evil and the imminent arrival of a savior descending from heaven. Such a person, seeing a strange aerial event, might interpret it as the approach of a threatening, punishing angel, or as the coming of a savior. However, in the hard-core cases of UFO reports we find no such thing; the witnesses frequently find their observations jarring to their own beliefs but insist nevertheless on what they have seen. Often those witnesses say that they never took UFO reports seriously or that they thought those reports were nonsense. When such a person sights a puzzling UFO, we would expect him to try very hard to categorize it in familiar ways. In fact I would find it puzzling and behaviorally anomalous if witnesses to a dramatic, ambiguous event promptly interpreted it in a way that lay outside their previous beliefs and contrary to the beliefs of others around them unless, indeed, their observations seemed quite unequivocal. This would be an extraordinary suspension of the usual laws of human behavior.

A few cases of so-called mass hysteria and hysterical contagion have been relatively well documented and described, such as the public reaction to Orson Welles's radio drama about an "invasion from Mars," the case of the phantom anaesthetist of Mattoon, the Seattle windshield-pitting epidemic, and the so-called "June Bug" epidemic in a southern factory.[3] Apparently the recipe for this type of hysterical outbreak is a combination of a high level of anxiety or tension with some kind of ambiguous event which is interpreted as posing a serious threat. The ambiguous event is transformed, in beliefs, into an unambiguously threatening event which apparently justifies the diffuse anxiety which was its antecedent.[4] The documented cases of hysterical contagion generally last a few days, or at most a few weeks. Welles's radio drama came at Halloween, 1938, a time of much anxiety concerning wars and invasions. Hitler's German armies had recently marched into Austria, and Japanese armies were advancing in China. The drama, realistically presented in the form of news bulletins and interviews concerning an alien spaceship landing in New Jersey, resulted in many kinds of hysterical actions, including thousands of panic-stricken phone calls, wildly fleeing automobiles, and impromptu shotgun brigades. The period of mass hysteria was brief, lasting a day or less. Around Seattle in March 1954, people began noticing pitting on their car windshields, and these observations were reported in the press. The reports closely followed news about the aftermath of H-bomb testing in the Pacific, including harm to Japanese fishermen and stories about the test getting out of control. Police received hundreds of calls, and public concern reached such a pitch that the Mayor made emergency appeals to the Governor and the President before detached study indicated that the pitting had nothing to do with H-bombs but was apparently normal road damage which drivers seldom notice. The windshield-pitting epidemic had two periods of activity: the first was scattered in several towns within about a fifty-mile radius of Seattle over a period of about a week; there was a pause of about one week; the second period was intense and concentrated in the city of Seattle for another week. In Mattoon, Illinois, there was an outbreak of mass hysteria involving reports of a mad gasser who was alleged to sneak around squirting gas at women. This outbreak was limited to the city of Mattoon and lasted about three weeks. In a southern factory in 1962 there was an outbreak of mysterious symptoms including nausea, skin rash, and fainting, which were attributed by the victims to the bites of tiny (indeed invisible) insects. Nearly all cases were in one area of the factory; 95 per cent of the cases occurred in a period of four days, and the entire set of reports covered about eleven days.

Some effort has been made to liken UFO reports to these cases of hysterical contagion. It appears quite clear that hysterical contagion contributes some cases to the massive number of reports, but there are many difficulties in trying to argue that the hard-core cases can be explained in this way. For one thing, the persons reporting UFO's frequently do not interpret them as serious personal threats. They often describe a UFO with puzzlement but not fear. For another, the continuation of UFO reports over at least decades and their spread over all parts of the world would both be unprecedented for a case of hysterical contagion. Also, the fact that many reports are made by people previously unfamiliar with UFO reports would argue against contagion as the mechanism underlying the best reports. And in many cases the events described by UFO witnesses are not fleeting and ambiguous events, such as the invisible insects and figures fleeing in the dark that have been described by witnesses in cases of hysterical contagion. Frequently they are accounts of prolonged observation with much solid detail. Finally, witnesses often report details which are consistent with other reports that have *not* been described in the mass media. Admittedly, it is difficult to establish a witness's lack of prior exposure to specific information. However, if the witness is not a UFO buff who reads special publications and if the news media have not reported the relevant details, then we are stretching a point to explain the reported details as the result of contagion.

It seems clear from the behavior of people who write about UFO's that there have come to be strong, socially supported systems of belief surrounding UFO reports. These systems of belief complicate the problem by interfering with perception and interpretation of events. Some UFO buffs, in writing case descriptions, load their reports with interpretation, making it difficult to separate fact from fiction. On the other hand, some skeptical scientists, faced with detailed reports by reliable witnesses, loudly and confidently assert interpretations which conflict strongly with available testimony and show a startling degree of disrespect for the reason and common sense of intelligent witnesses.

## Hard-Core Cases: Physical Event or Motivated Misperception?

Let us grant that many UFO reports are misidentifications of familiar objects, perhaps given a boost sometimes by such processes as psychological projection and hysterical contagion. The question remains whether there is a residual subset of UFO reports for which there must have been a real, novel physical stimulus, or whether it is plausible to argue that the "hard-core" cases are also systematic misperceptions, guided by psychological mechanisms such as projection and contagion of belief.

Let us consider first the question of the credibility of human testimony. Our legal system is based largely on the assumption that, under certain conditions, we can accept human testimony as factual. Many people, including attorneys and judges as well as behavioral scientists, have rather clear-cut criteria for assessing the credibility of testimony: the witness's reputation in his community, previous familiarity with the events and persons involved in the testimony, apparent motives for prevarication or distortion, and internal characteristics of the testimony such as consistency, recency, verifiable detail, and so forth. Also, testimony is more credible with multiple witnesses, especially independent ones, and with multiple channels of observations (e.g., both visual and auditory; both unaided observation and observation through instruments). If we apply these criteria to the witnesses and the testimony of hard-core UFO reports, some of them stand up better than many a court case. In some cases there has even been a kind of "cross-examining" of witnesses in reinterviewing by scientists such as Hynek and McDonald. Examples of hard-core cases in which I find familiar explanations, including systematic misperception, implausible are the Lakenheath case, reported in the Colorado Report and more fully by McDonald in the present symposium; and the RB-47 case, reported in the Colorado report and in much more detail, with additional witnesses, by McDonald in the present symposium.[5] Other examples of hard-core cases include Hollywood on February 5–6, 1960; Arrey, New Mexico, on April 24, 1949; Red Bluff, California, on August 13, 1960; Admiralty Bay on March 16, 1961; and Redlands, California on February 2, 1968.[6]

One problem in assessing the testimony in such cases is the difficulty in establishing whether witnesses did, in fact, report independently or whether they were in a position to influence one another's reports. Another problem is that of determining the pre-existing knowledge and belief of a witness. There are many cases in which witnesses deny previous knowledge and cases in which they strongly deny ever believing reports of UFO's before they saw one. Nevertheless, human memory is fallible in such matters, and it is conceivable that witnesses are unconsciously influenced by information read or heard long before.

I believe that most behavioral scientists who examine the evidence would agree that reports as persistent and patterned as hard-core UFO reports must be systematically motivated in some way, not simply random misperceptions. Either there must be a distinctive physical phenomenon which these witnesses have observed, or there must be a powerful and poorly understood motivation rooted in projection, or contagion of belief, or a similar mechanism. Given these alternatives, I find it more plausible to believe that there is a distinctive physical stimulus than to believe that multiple witnesses misperceive in such a way as to make them firmly believe they saw something which jars their own beliefs and subjects them to ridicule of their associates—something they sometimes report observing both with unaided eyesight and through instruments over a prolonged period, and something they can describe calmly and in detail.

## Scientists' Responses to Off-Beat Phenomena

In our scientific ideals we like to set goals for ourselves and our students that are superhuman in their detachment and openness to challenge and revision. In the hard world of real scientists, there are altogether too many anecdotes which suggest that scientists, too, are human. When Galileo's telescope made it possible to sight the moons of Jupiter, many refused to look through the telescope.[7] They "knew" that there could not be such bodies around Jupiter, and therefore they "knew" that the telescope was a deceptive instrument. Even more instructive cases come from the history of meteorites. To quote an account by C. P. Olivier:

> In the next three centuries [after 1492], a good many meteorites fell in Europe, but the reaction against superstitions of the Middle Ages led the scientists of the day to such great skepticism that they refused to face facts, in some cases. Perhaps the most notorious instance refers to meteorites: in the 18th century the learned men of the day did not *believe* stones could fall from the skies, hence they affirmed they did not. Even the great French Académie des Sciences went on record denying that meteorites had an origin outside the atmosphere, despite accounts of falls by reliable witnesses, which were ridiculed, and the splendid pioneer work of Ernest F. F. Chladni about 1794.[8]

It has been reported [9] that skepticism was so strong that the reports of witnesses were changed to conform with acknowledged theories, and museum keepers followed scientific advice and threw away meteorites lest they be accused of clinging to foolish superstitions.

There are many anecdotes about the reluctance of scientists, often distinguished ones, to accept new observations. The point seems to be that scientists are human and behave according to the same principles of human behavior as nonscientists. Indeed we might describe the body of scientific knowledge accepted at any given time and the people who bear that knowledge as constituting an unusually strong belief system which resists inconsistent items of knowledge even more powerfully than a layman defending his political beliefs.

To the extent that observation challenges established beliefs, scientists resist accepting the observation. This resistance seems to take several forms: One form is the avoidance or denial of evidence, as seen in those who would not look through Galileo's telescope or those who refused to believe reports of meteorite showers. Another form of resistance shows up in illogical arguments by men who are customarily precise and logical. For example, we find some scientists arguing something like this: "I can cite hundreds of cases of people who were excited and reported an aircraft or a star as a UFO and hundreds of humorous cases of unbalanced people with demonstrably false stories; *therefore* it is plausible that the rest of the cases are similar." I know from personal experience as a military flyer in wartime that flyers sometimes shoot at Venus or at an island, believing it to be an aircraft. It would be foolish for me to conclude from this that there were no aircraft in the sky. Another form of avoidance is the kind of buck-passing that has occurred often with respect to UFO reports. If there *is* a new physical phenomenon behaving as the reports describe, this may force physical scientists to confront an anomaly and modify something in their present knowledge and belief to accord with these observations. Consequently they say that there is no physical phenomenon; it is all psychological: human errors of observation and interpretation, mental aberrations, hysterical contagion, and the like. On the other hand, if there is *not* a physical phenomenon, then behavioral scientists are confronted with an anomaly and may have to modify something in their knowledge and belief to account reasonably for the persistence of so many apparently sound UFO reports. Consequently, I, speaking as a behavioral scientist, say that there *must* be a real physical phenomenon. So we pass the buck back and forth without forming *any* adequate explanation, either physical or behavioral.

The very strength of our resistance to the evidence on UFO's suggests to me that there is clearly a phenomenon of surpassing importance here. It is going to force some of us to make some fundamental changes in our knowledge, and this is a good definition of scientific importance. The arguments are really arguments about *who* has to change. In whose domain does this phenomenon lie? Do the physical scientists have to accept the existence of such a puzzling and anomalous physical object or phenomenon? If so, they must set out to account for it. Or do the behavioral scientists have to accept the puzzling and anomalous fact that hundreds of intelligent, responsible witnesses *can* continue to be wrong for many years? If so they must then set out to account for this massive fallibility.

## NOTES

1. T. Shibutani, *Improvised News: A Sociological Study of Rumor* (Indianapolis and New York: Bobbs-Merrill, 1966) pp. 32–34.

2. R. B. Zajonc, "Cognitive Theories in Social Psychology," in *The Handbook of Social Psychology,* ed. by G. Lindzey and E. Aronson, Vol. I (2d ed.; Reading, Mass: Addison-Wesley, 1968), Chap. 5, pp. 320–411; W. J. McGuire, "The Nature of Attitudes and Attitude Change," in *The Handbook of Social Psychology,* Vol. III (1969), Chap. 21, pp. 136–314; H. Tajfel, "Social and Cultural Factors in Perception," in *The Handbook of Social Psychology,* Vol. III (1969), Chap. 22, pp. 315–394; R. P. Abelson *et al.*, eds., *Theories of Cognitive Consistency: A Sourcebook* (Chicago: Rand McNally, 1968).

3. H. Cantril, *The Invasion from Mars* (Princeton: Princeton University Press, 1940); D. M. Johnston, "The 'Phantom Anaesthetist' of Mattoon: A Field Study of Mass Hysteria," *Journal of Abnormal and Social Psychology* 40 (1945): 175–186; N. Z. Medalia and O. N. Larsen, "Diffusion and Belief in a Collective Delusion: The Seattle Windshield Pitting Epidemic," *American Sociological Review* 23 (1958): 180–186; A. C. Kerckhoff and K. Back, *The June Bug: A Study of Hysterical Contagion* (New York: Appleton-Century-Crofts, 1968).

4. N. Smelser, *Theory of Collective Behavior* (New York: Free Press of Glencoe, 1963), Chaps. 5 and 6.

5. E. U. Condon, *Scientific Study of Unidentified Flying Objects* (New York: Bantam Books, 1969), pp. 163–164; 248–256; 136–139; and 260–266.

6. *Symposium on Unidentified Flying Objects,* Hearings before the Committee on Science and Astronautics, U.S. House of Representatives, 90th Congress, 2d Session, July 29, 1968 (Washington, D.C.: U.S. Government Printing Office, 1968) pp. 54–57; 63–64; 109–110; 64–65; 52–53.

7. B. Russell, *The Impact of Science on Society* (New York: Simon and Schuster, 1953).

8. C. P. Olivier, "Meteor," in *Encyclopedia Americana,* 1965 ed., Vol. 18, p. 713.

9. F. A. Paneth and M. H. Hey, "Meteorites," in *Encyclopaedia Brittanica,* 1970 ed., Vol. 15, pp. 276–277.

# 10

## Psychology and Epistemology of UFO Interpretations

DOUGLASS R. PRICE-WILLIAMS

This discussion is concerned with the psychology and epistemology of *interpretations* given to *reports* of unidentified flying objects, and not with the individual make-up of (or logic used by) witnesses. My theme is that distinctions must be made between description, definition, and explanation. Failure to respect these distinctions frequently leads to lack of clarity in discussing the reports, and often to logical mistakes. I propose to enumerate four stages through which inquiry has to follow, and to comment in passing on attempts made so far.

*Stage I.* Reports of curious aerial (for the most part) phenomena are generated. This is the starting point and here already we meet the first difficulty. Our primary descriptive term and the title of this symposium has drawn logical complaint from people of quite differing interpretations: Hynek,[1] Page,[2] Baker, Menzel,[3] and Vallee.[4] Each of the three words runs into trouble: "unidentified" because it embraces too much; "flying" because it suggests something mechanistic (we do not talk of a cloud flying, except in poetry); "objects" because it already presumes a conclusion. Smuggled into this term is already an assumption masquerading as a description. The error is compounded when the term "flying saucers" is regarded as synonymous with "UFO's," as now we have an *explanation* masquerading as a description. However, it looks as though we are stuck with the term "UFO's," and although something like "anomalous observational phenomena" (Baker's term, Chapter 8, above) is logically preferable, we had better go on using "UFO's," remembering that the usage does not commit us to any interpretation.

*Stage II.* The reports now undergo differentiation. Investigators—experts of various kinds—have managed to eliminate the "unidentifiableness" of many reports, indeed the majority, and have traced them to quite identifiable and understood phenomena. Nevertheless there is left a residue of still curious and puzzling phenomena. It is at this second stage that controversy really begins. What appears to be an insignificant and unrelated residue, to one viewpoint, is to an opposing viewpoint both significant and related. Whereas the first group may dismiss the residue as something akin to error variance, the second group accept the residue as signal, or at least are committed to the viewpoint that it *could* be signal, for which further investigation is necessary. Such further investigation brings us to the next stage, but there exists a problem which impedes easy transition to the third stage. This is the problem of *populations* of reports. It becomes clear very quickly to the student of the subject that different authors are often alluding to different samples, and it is by no means obvious how the different samples are related. It would seem that three populations can be organized.

*Population A* comprises those reports that are explained by reference to known phenomena, reports which our questioning second group is not prepared to defend as still requiring investigation. In other words, everyone agrees that Report X is that of a meteor, so let us eliminate it from our residue. No controversy here.

*Population B* includes those reports that are explained by the first group of investigators as exemplars of known phenomena but whose explanation meets disagreement from the second group. There is valid and real controversy here, but note that, at this Stage II, we need not necessarily have a conflict of two hypotheses, two kinds of explanations. Someone may disagree with Report Z as constituting ball lightning without being committed to an explanation in terms of extraterrestrial vehicles. The controversy is over whether one should leave Report Z in the residue of cases that still need investigation or not. The reverse of this situation can be noted in the Colorado Report. Rejection of a photograph of a UFO as constituting evidence of an extraterrestrial vehicle does not commit the rejecter to an explanation in terms of known phenomena.[5]

*Population C* covers those reports which both groups agree are unidentifiable and unexplainable in terms of known phenomena. Presumably all the "unidentifieds" in the Air Force files belong to this population, as well as the unexplainables in the Colorado Report.

All three populations are defined in terms of having been examined and evaluated. It becomes clear that there are many reports which have not been introduced to the scientific forum at all and which, therefore, await assignment. It is necessary to make what may appear to be a somewhat trite point about population grouping, as otherwise we can never be certain whether we are talking about the same thing.

It is worthwhile noting that the Colorado Report essentially terminates at Stage II. Their scientific filtering still left between 20 and 30 per cent of the reports unexplained. The report's reluctance to proceed further apparently stems from the decision to test the hypothesis of extraterrestrial intelligence on the complete sample, as the majority of cases could be explained otherwise. A consequence of this approach is to present equivocal conclusions when unexplained instances are confronted. Thus, regarding photographic cases: "The present data are compatible with, but do not establish either the hypothesis that (1) the entire UFO phenomenon is a product of misidentification, poor reporting, and fabrication, or that (2) a very small part of the UFO phenomenon involves extraordinary events." [6] Again, with Colorado Case No. 14, First Sighting: "There is no reason to doubt the credibility of the sighting; however, the question of *what* was seen remains unresolved." [7] Or again with Case No. 17: "Investigation revealed neither a natural explanation to account for the sighting, nor sufficient evidence to sustain an unconventional hypothesis." [8] It would seem that the Colorado investigators had difficulty in distinguishing (*a*) phenomena of a certain kind that are unexplained, from (*b*) phenomena that could be attributed to extraterrestrial intelligence. As the only hypothetical properties of the latter appear to be those which are reported as the former, the hypothesis of ETI formulated in this way leads only to circular reasoning, and is not in a form amenable to empirical acceptance or rejection.

*Stage III.* This, then, is the stage where we have for inspection at least Population C, and those reports from Population B which, after further debate, may get transferred to C. It must be remembered that the residue of reports at this stage only have the communality of still being unexplained. That they may have *descriptive identity* or *class definition* has yet to be demonstrated and argued, and cannot be assumed. Descriptive identity is not to be misunderstood as explanatory identity. A null-set can still have properties. We can make classes of disease symptoms without knowing their causes. Nevertheless, Stage III is a key link in our sequence, as failure to define the data at this point makes further analysis unamenable to systematic investigation. We need, therefore, to give considerable thought to it.

The problem is to extract information from the reports, regardless of whether a physical, psychological, or sociological type of explanation is to be invoked.* Now information is reduced by the presence of equivocation and by the presence of noise. We need to apply these concepts from information theory to the case in hand. In doing so, I am concerned only with the major source of data, that is to say human testimony. The same principles must apply to other sources, such as information from radar and automatic instruments, but these contain different technical factors, and others in this symposium are concerned with them.

Class definitions have been attempted previously, by NICAP [9] and by Vallee.[10] Both are mainly based on the principle of dividing the descriptions into the main variables or attributes of shape, size, color, kinemat-

* Dependent on the type of explanation, however, the information input may vary. A sociological inquiry would require information about current knowledge of UFO's in the area concerned, in the form of recent motion pictures, newspaper articles, and so on. A psychological inquiry might focus on archetypal themes quite apart from descriptions of the phenomena as given by witnesses; this approach has been taken by Jung, *Flying Saucers: A Modern Myth of Things Seen in the Skies* (New York: Signet Books, 1969, originally published in German in 1958), by John Michell, *The Flying Saucer Vision* (New York: Ace Books, 1967), and by Jacques Vallee, *Passport to Magonia: From Folklore to Flying Saucers* (Chicago: Henry Regnery Co., 1969). Attention to depth psychological material does not necessarily resolve the prime enigma, for we are left with the option of concluding either that the entire phenomenon can be explained by not altogether well-understood mechanisms of neurology and analytical psychology, or that the phenomenon is indeed "true," or a third possibility that Jung himself tentatively suggested: "that UFO's are real material phenomena of an unknown nature . . . [on which] unconscious contents have projected themselves and given them a significance they in no way deserve" (pp. 117–118 of Signet edition).

ics, and others on the one hand, and on witness reliability on the other. Vallee, in particular, is sensitive to some aspects of the noise factor, distinguishing his types to show resemblance or lack of resemblance to interfering irrelevant stimuli, such as satellites or meteors. At the outset it should be obvious that the data we are inspecting and analyzing are those of the report of the witness(es). As McDonald has shown, secondary elaborations, such as newspaper accounts (which have their own motivation, e.g., interest to their readers), are often untrustworthy.[11] Reduction of noise must include this factor. Also, there is the noise factor of time delay in reporting the event. Hall has covered other aspects of witness reliability.[12] On the whole, noise factors appear to be well appreciated in this field.

Equivocation factors are not so obviously appreciated. Before we are confident of assigning percentages to attributes of shape, size, and so forth, and basing explanations on them, we should appreciate that descriptions are peculiarly open to assumptive contexts. Shepard understands this,[13] and his remarks refer to fresh methods of information retrieval for reports: a very useful suggestion which, however, means starting anew. Wertheimer undoubtedly appreciates this,[14] but his short chapter in the Colorado Report concludes only that there is room for error. We need to find a method of reducing equivocation without going to the extreme of throwing up our hands in horror at the fallibility of human testimony, thereby invoking McDonald's complaint that the psychologists' "puristic insistence on the miserable observing equipment with which the human species is cursed makes me wonder how they dare cross a busy traffic intersection." [15]

Equivocation can be regarded as information that the organism cannot discriminate reliably, and this is the very heart of our difficulty in definition. When a witness reports that he has seen for a short time a disk-shaped object hovering one hundred feet away from him at tree-top level, what credibility can be placed on the descriptive attributes of disk-shaped, one hundred feet, hovering, and tree-top level? There is little doubt that the difficulties are formidable. Most people are unused to angular estimation; most people tend to express themselves in thing-language and not process-language. Many reports simply do not give the basic information necessary to judge the relative accuracy of estimates of shape, size, distance, and so on. Perceptual cues that may or may not have been present to the witness are often not noted in the reports. We often are not told what visual angle the observer made his observation from; what degree of illumination was present; what frame of reference was used. We do not know whether the witness has some implicit assumptions as to shape and size of UFO's. The list could be extended.

There are only two ways of adjusting to this state of affairs, other than planning anew the entire retrieval procedure of reporting these phenomena in order to accommodate to the facts of visual perception. The first would be to go painstakingly through existing reports, noting what aspects of what reports can probably be relied on. We can note whether the "object" was seen against a background which contains what has been called "microstructure" or against a kind of background like the sky which contains "film-color." [16] This is a basic datum, for judgment of distance and hence size depend on it. We can note whether the phenomenon was observed from directly above or below or viewed obliquely and how this correlates with the reported shape.[17] We can ask whether there are quite different reports of movements of the phenomena when the witness is relatively stationary as against when he is moving as in an airplane or automobile, or, alternatively, whether they are much the same. Or again, whether the reported motion was seen against a fixed background or against a possible moving background, like a cloud. Such, and further, probing of the data could be made, and would constitute some kind of check, however rough.

The second way is to examine the data for what might be called latent descriptions. A latent description denotes a relationship between attributes which emerge as a statistical invariant across the mass of reports. This requires going beyond the case-by-case approach and necessitates a cross-correlation of a number of reports, on a statistical basis. NICAP's [18] linkage of reported motion and color is a step in this direction, as is the Vallees' analysis of estimated size of object to distance from the observer,[19] but the analysis needs to be done on a far larger scale than has been tentatively attempted hitherto. This could be done separately for Population A and Population C reports, so that comparative assessment is possible. Unfortunately, it appears that the mass of data has not even been marshalled into a form which lends itself to this kind of analysis.* Interpretations of such relationships, if they prove to exist, is another matter, of course, and this brings us to the last stage.

*Stage IV.* Having arranged and refined the data, the stage is relatively clear for hypotheses to be brought to bear on them. Our sequence of stages is not meant to insinuate an inductive approach in which the systematic refinement of data astonishingly reveals an explanation. A hypothesis may demand a datum which nobody has yet reported or no one thought of asking for, and may include factors other than descriptions. The intention here is only to seek a confrontation of data and hypotheses, a feature somewhat lacking in this field. This is due partly to the "dirty" condition of the data and partly to the dominance of *a priori* assumptions. Indeed, assumptions have tended to be voiced in ways which encourage ignoring of the data rather than relating the data to various hypotheses. Furthermore, little thought has been given to how certain hypotheses could, in principle, be tested. This is crucially true of the extraterrestrial hypothesis. I have already indicated that "proof" of this cannot rest merely on the bizarre nature of the reports. Apart from trapping such an object, the only approach that seems at all possible is to posit a model embodying aerodynamic and engineering properties that are then matched against the observed data as reported. This approach has been tried by some,[20] yet there are logical hazards even here. One might say there is a higher and lower limit within which such

* Project Bluebook Special Report No. 14, 1955, made a study in which the data up to 1952 were converted into IBM punched-card form. The analysis, however, consisted of nominal classifications of attributes and geographical and time distributions plus single attribute comparison with known phenomena. While not without value, the study did not pursue the line of investigation demanded here. Unfortunately, we learn from D. Saunders in his book *UFOs? Yes!* (New York: Signet Books, 1968, p. 115) that the card deck was thrown away. J. A. Hynek (in a letter reported by W. Markowitz, "The Physics and Metaphysics of Unidentified Flying Objects," *Science* 157 (1967): 1276) reveals that the Air Force does not have the material in machine-readable form. D. J. Pearson (then at the Centre for Computing and Automation, Imperial College, University of London) did perform a time-series analysis of 1,500 British sightings. He was prepared to construct a master program using the Colorado Project's data, when communication with the project staff broke down (*Flying Saucer Review* 14, (1968): 28). The negotiations with the Colorado Project involved 7,000 reports. After Dr. Saunders left the project, the data were not allowed to be released to Pearson (Personal communication to me from Pearson, Oct. 21, 1969).

a model can operate. The higher limit is bounded by postulating "magical" mechanisms, with which anything is possible. The lower limit is bounded by the fact that if the model makes physical sense, then presumably it has the status of an invention, and we could build such a thing. It would be truly ironic if a "flying saucer" were constructed on the basis of clues that were wholly the product of a twentieth-century myth.

Unfortunately there is a poverty of hypotheses between the extremes of extraterrestrial machines and misinterpretations of known phenomena. There is no doubt that the peculiar nature of the data and the main source of their generation present genuine and difficult epistemological and methodological problems. What we have is a compound relationship of data, different hypotheses, observational uncertainty, and reliability of witnesses. Possibly what we need primarily is a model for sorting out these factors, probably some application of the Bayesian model.[21]

## NOTES


1. J. A. Hynek, "The Condon Report and UFO's," *Bulletin of the Atomic Scientists* 25 (1969): 40.
2. T. Page quoted by D. H. Menzel and L. G. Boyd in *The World of Flying Saucers* (New York: Doubleday, 1963), p. 143.
3. R. M. L. Baker, Jr., in *Symposium on Unidentified Flying Objects,* Hearings before the Committee on Science and Astronautics, U.S. House of Representatives, 90th Congress, 2d Session, July 29, 1968 (Washington, D.C.: U.S. Government Printing Office, 1968), p. 126; D. H. Menzel in *Symposium,* p. 202.
4. J. Vallee, *Anatomy of a Phenomenon* (Chicago: 1965), pp. 101–103.
5. W. K. Hartmann in E. U. Condon, *Scientific Study of Unidentified Flying Objects* (New York: Bantam Books, 1969), Section III, Chap. 2, pp. 75–76.
6. *Ibid.,* p. 86.
7. R. J. Low *et al.* in Condon, *Scientific Study,* Section IV, Chap. 2, p. 287.
8. J. E. Wadsworth in Condon, *Scientific Study,* Section IV, Chap. 2, p. 295.

9. R. H. Hall, *The UFO Evidence* (Washington, D.C.: NICAP, 1964), Section XII, pp. 143–171.

10. J. Vallee, *Anatomy*, pp. 103–108.

11. J. E. McDonald in T. Bloecher, *Report on the UFO Wave of 1947* (privately printed, 1967), Introduction, pp. iii–xi.

12. R. L. Hall in *Symposium on Unidentified Flying Objects*, p. 101.

13. R. N. Shepard in *Symposium on Unidentified Flying Objects*, pp. 225–233.

14. M. Wertheimer in Condon, *Scientific Study*, Section VI, Chap. 1, pp. 559–567.

15. J. E. McDonald in *Symposium on Unidentified Flying Objects*, p. 38.

16. J. J. Gibson, "Perception of Distance and Space in the Open Air," in D. C. Beardslee and M. Wertheimer, eds., *Readings in Perception* (Princeton: Van Nostrand, 1965), p. 423.

17. See A. Durham and K. Watkins, "Visual Perception of UFO's, Part II," *Flying Saucer Review* 13 (1967): 25.

18. Hall, ed., *The UFO Evidence*, p. 152.

19. J. and J. Vallee, *Challenge to Science* (New York: Ace Books, 1966), p. 186. The allusion in the book is really to size constancy.

20. For example, R. H. B. Winder, "Design for a Flying Saucer," *Flying Saucer Review* 12 (1966), 13 (1967); Les E. Huntley, "UFO: A Possible Method of Propulsion," *APRO Newsletter* 3 (1969); Leonard G. Cramp, *Piece for a Jig-Saw* (Cowes, Isle of Wight: Somerton Publishing Co., 1966); Lieutenant Plantier, in Aimé Michel, *The Truth about Flying Saucers* (London: Transworld Publishers, Corgi Book, 1958), Part 3, Chap. 2.

21. For the basic application of Bayes's theorem to psychology, see W. Edwards *et al.*, "Bayesian Statistic Inference for Psychological Research," *Psychological Review* 70 (1963): 193–242. For application to inferences about data-generating hypotheses, see C. F. Gettys and T. A. Willke, "The Application of Bayes's Theorem when the True Data State Is Unknown," *Organizational Behavior and Human Performance* 4 (1969): 125–141. In relation to the problem of weighing evidence, see I. J. Good, *Probability and the Weighing of Evidence* (London: Griffin and Co., 1950), especially Chap. 6, pp. 62–75.

# 11

## Psychiatry and UFO Reports

LESTER GRINSPOON
and ALAN D. PERSKY

As psychiatrists we make no pretense of being able to resolve any of the central questions raised during the past few decades about "unidentified flying objects." There is no question that some "things" have been observed, and described, by a variety of observers, both singly and in groups. Speculation about the origin and nature of what has been observed has provoked a controversy which at times appears to be an emotional donnybrook. There is little disagreement that no unifying answer is available. Rather, one deals with a diverse group of phenomena, facts and fabrications, the real and the imagined, the natural and the "unnatural," all of which have been lumped together under the now highly cathected term, "UFO's." The degree of controversy surrounding the subject suggests an extraordinary appeal, an emotionalism far exceeding the usual involvement in scientific subjects in our scientific age. Such fervor is more commonly observed when matters of politics or religion and morality are involved. One has only to consider the worldwide excitement lasting more than twenty years, the detailed investigations into the subject, and the heat of the opposing arguments to be suspicious that, whatever is being seen, there is more here than meets the eye.

We will limit ourselves to consideration of mental processes, as they occur in individuals, and not attempt to deal with the equally complex group phenomena, forms of mass hysteria, and epidemiologic considerations, nor with physiologically determined visual distortions. To paraphrase Freud we will deal with the "Psychopathology of Everyday Unidentified Flying Objects," with some excursions into the distortions of the mentally ill, particularly the psychotic, the borderline, and the personality disorder. It is clearly beyond our scope to consider what has been observed in more than a small fraction of cases. If we then contribute to the comprehension of even a small number of cases, we will have succeeded in reducing the aura of mystery and intrigue that surrounds the subject in such a persistent fashion.

Fundamental to our discussion is the fact that there exist unconscious mental processes which are of paramount importance in the conscious functioning of all men at all times. Furthermore, as has been amply demonstrated, the expression of unconscious processes (whether in persons termed healthy or sick) is often in the form of symbols. This symbolism transcends the individual, the race, the culture, and even the historical era. Studies by sociologists, anthropologists, and psychoanalysts indicate that man's psychological nature is such that he first attempts to explain the world by magical or animistic beliefs and then by religious, and finally by scientific means. This is both a phylogenetic and ontogenetic process. It is at once both the history of man and of a man: the developing child psychologically recapitulates this history as he matures. The stages are not mutually exclusive, the development is rarely complete, and in fact the progression does not occur without vestiges of earlier stages persisting in some form. Even the most sophisticated among us has at the very least a potential for vestigial modes of thinking.

In times of severe stress, either from the environment or from intrapsychic conflict, man reverts or regresses, falls back on more primal modes of thinking. Often when the challenge is severe, when science and religion are undeveloped, unavailable, or seem of little assistance, then the magical and the mythical explanation is utilized. More sophisticated thought processes fail, and what is commonly referred to as primary-process thinking occurs. Primary-process thinking is observed directly and inferentially, the former in the very young, in the very ill, and under the stimulus of psychoanalytic treatment. The laws that govern the unconscious differ markedly from what is familiar to conscious thought. Primary-process thinking is characterized by the absence of negatives, conditionals, or other qualifiers, and by the use of allusion, analogy, displacement, condensation, and symbolic representation. It is the source of myth, of magic, of fantasy.

Its relevance to this discussion exists where a significant degree of primary-process thinking affects the observer's sighting of a UFO and prevents him from making an objective report. In just how many of the UFO reports these factors play a part is, of course, unknown. We suspect, however, that a significant fraction of the observations may be involved, for several reasons. Problems of methodology in collecting these data notwithstanding, it seems clear that there is a high prevalence of untreated as well as treated cases of psychological disorder in the community. Considering seven of the most recent studies, ones in which the data were gathered through interviews with the subjects, the range of psychological disorder varies greatly from 13.2 per cent to 64 per cent.[1] As one looks at the results of these investigations over a period of time, there is a clear tendency for rates to be higher in more recent studies. Especially notable is the increase in the 1950's and the 1960's. It is premature to infer a true change in prevalence over time. The change is most likely due to methodological differences in the studies. Nonetheless, it cannot be ruled out that this increase is to some extent a consequence of the increasingly anxious times in which we live. In addition to those in the community who have more or less persistent psychopathology, there are some at any one time whose mental functioning is transiently disturbed by a stressful situation, most commonly an important loss or a serious threat to well-being. In a study of reactions to President Kennedy's assassination,[2] interviews were done five to nine days after the assassination. Eighty-nine per cent of the sample reported experiencing physical and emotional symptoms during the first four days, and 50 per cent reported they still had at least one symptom at the time of the interview.

Furthermore, there is, among those whose grasp of reality is already somewhat shaky, a tendency to invest more in fantasies. People with more or less impaired ego function, be it fairly permanent or transient, are more likely to exhibit primary-process thinking and be attracted to magical phenomena. Thus we are conjecturing that a population of emotionally disturbed people is more likely than a nondisturbed population both to be attracted to the possibility that we are being visited by extraterrestrial forms of intelligent life and to make observations that support this possibility.

Among the disorders of perception and thought which may be involved in misrepresenting the reality are illusions, delusions, and hallucinations. The conversion of sensory impulses into an accurate sensory image is dependent on an ego whose function is intact, particularly its synthetic function. One of the ways in which even moderate degrees of emotional disturbance may affect mental function involves some degree of incapacitation of the ego in this regard. This may lead, through perceptual misinterpretation of sensory images, to the formation of an illusion. Particularly likely to lead to this kind of disruption are urgent drives and impulses, overwhelming wishes, and toxic, febrile, and intense affective states. Under certain conditions, misinterpretations may occur that reflect some affect or express some wish or drive. For example, a person engaged in a fantasy over which he feels a great deal of guilt may be startled by an otherwise unobtrusive noise as he misinterprets it as the intrusion of a reproaching figure into his fantasy. Marked expectation, fatigue, drugs, guilt, anxiety, or fear predispose to illusional interpretations. In mental health, but particularly in mental disorder, the emotional life imbues and tends to influence perceptual experience according to the needs of the person.

In the case of an illusion, an image symbol of a real object is formed, but for psychological reasons it is misinterpreted, whereas hallucinations are regarded as perceptions which occur when there is no impulse created by the stimulation of a receptor. While it is then a perception without object, it nonetheless constitutes an actual part of the subject's mental life. An ego which is seriously disordered, either because of a severe mental illness or a toxic state, allows the breakthrough of preconscious or unconscious material into the consciousness in the form of sensory images in response to psychological needs. These hallucinated images, which the subject accepts as reality, often represent the projection onto the outer world of such inner needs as wish-fulfillment, enhancement of self-esteem, censure, a sense of guilt, or self-punishment. Thus, they provide satisfaction of repressed and rejected impulses toward the goal of achieving a more satisfying reality. In the mentally healthy person, most perceptions produced by casual stimuli from the environment are ignored. The content of perceptions in hallucinations, however, is often so intimately subjective that reality must be made to harmonize with it. The functional capacity of the ego for testing any reality that does not harmonize with the hallucination is often suspended. Visual hallucinations occur less frequently than auditory hallucinations and are observed most typically in people with acute infectious diseases or toxic reactions such as those induced by the ever increasing number of drug experimenters.

While we are all prone to generate anxiety-relieving or other psychologically useful fictions, in some the necessity to satisfy urgent inner needs is so overwhelming that the ego gives up some of its claim on reality, and delusional ideas make their appearance. Reality is reshaped to make it synchronous with the emotional needs of the delusional person. But it should be emphasized that by "delusional" we are not referring to just any type of false belief that may be a function of inadequate intelligence, education, or experience. Rather we are considering the incorrect understanding and use of facts and evidence in response to some inner needs of the personality. In other words, the important determinant in the creation of a delusion is affective rather than ideative or cognitive. Such delusions are often attempts to deal with the special difficulties and stresses of a life-situation through the substitution of fantasy in an effort to supply what the reality has denied. They differ from healthy beliefs and fantasies, however, in their irrational fixity even in the face of what would be considered incontrovertible evidence to the contrary.

Sources of these difficulties may be found in intrapsychic conflicts which require defense against anxiety, in frustrated drives and hopes, and in feelings of inferiority, inadequacy, rejection, or guilt. A young psychiatric social worker who was suffering from acute feelings of anxiety and loss, as she terminated her psychoanalysis, developed an elaborate fixed delusional system wherein another psychoanalyst at the hospital where she worked had agreed to divorce his wife and run off with her. In this instance, the function of the delusion is transparent; it replaces her analyst with another one, but now in a relationship with a promise of more permanence. Often, however, the function is far from obvious and its interpretation remains speculative. This is not to say that the delusion is without significance or purpose, rather that its adaptive value is concealed by its symbolically disguised content. When the central theme of the delusion is developed and conclusions are so logically deduced from the premises assumed that a coherent and connected organization of ideas is established, then the delusion is said to be systematized. If the delusional system does not spill into other areas of the subject's life, then it is considered to be encapsulated.

Of the several categories of mental disturbance from the ranks of which one might expect unreliable reports of UFO sightings, ambulatory schizophrenia would be expected to be most important. People who suffer from this illness frequently have hallucinations, although more commonly auditory than visual, and often are delusional. One would expect that a grossly psychotic reporter would be recognized as such by the person accepting the report in a vis-à-vis situation. Of course, this is not always the case, as it depends on the sophistication and experience of the person who is evaluating the reporter. Recognition of psychosis may be especially difficult if the report is part of a relatively fixed, systematized, and encapsulated delusion. Here, aside from the delusional system there is no other apparent disturbance of thought processes, nor is there evidence of disturbance of affect, and the reporter continues to function normally in the other areas of his life. The existence of an underlying psychosis may be especially difficult to recognize if all that is available is a written account of the UFO sighting or experience. In fact, in this instance even the most experienced clinician would find it almost impossible to make a diagnosis from many if not most reports of this nature.

There is an unusual type of psychosis which may be pertinent to this discussion. It is a communicated form of mental disorder, known as *folie à deux*. This is a psychosis in which one of two intimately associated people develops certain mental symptoms, particularly delusions which are communicated to and accepted by the second person. This dual psychosis usually involves a parent and child, two siblings, or a husband and wife. The person suffering from the primary psychosis is the dominant individual, while the one who develops the induced psychosis is of a submissive and suggestible type, dependent upon and having a close emotional attachment to the infector. The primary psychotic may have at first a rather limited delusion which, as he develops it, systematizes it, and invests more and more in it, he imposes on the weaker person, who comes to share and even participate in the development of the systematized delusional ideas of the dominant person. We mention this type of psychosis because we believe that it may be offered as an alternative hypothesis to the given understanding of one of the most popularized UFO reports, which ethical and legal considerations prevent us from citing.

An important group of psychopathological constellations which have in common a rather stable form of pathological ego structure are the borderline personalities. With ego pathology which differs from that found in the neuroses and the less severe characterological illnesses on the one hand and the psychoses on the other, these people are considered to occupy a borderline area between neuroses and psychoses. However, transient psychotic episodes may develop in people with borderline personality organization particularly when they are under severe stress or the influence of alcohol or other drugs. Thinking disturbances of these people, as a rule, do not attract much attention since they are not as pronounced as in the psychoses. However, there is almost invariably a disturbance of judgment, of connecting cause and effect, and particularly of what is called "common sense." Their faculties for observation and memory are often poor. These failures are selective and emotionally determined. Because they are prone to distort reality, like the ambulatory schizophrenics, a population of borderline individuals could be expected to generate more UFO reports than a healthier population. Furthermore, with the tendency of some borderline personalities to become fanatic about particular issues, especially ones which allow for the expression of paranoid thoughts, one might expect that there would be those from this group who tenaciously cathect an extraterrestrial hypothesis of unidentified flying objects.

We have already mentioned that even relatively healthy people during times of acutely experienced stress may develop some degree of personality disruption. Again, we cannot be certain that such a group of people would be more susceptible to illusion formation, but in view of the tension, anxiety, and fear that they experience, we suspect that they are more prone at this time than during nonstressful periods. These disorders are usually short-lived, and if it is true that people who suffer from them may be more susceptible to a UFO illusion, it would be our guess that such a phenomenon would not become a highly cathected part of such a person's experience. We would expect that as he recovered from the transient personality disorder he would find that he has an increasing capacity to reality-test the illusory experience and decreasing intrapsychic need to cathect it.

Finally there are the antisocial or psychopathic personalities. Whereas in the above-mentioned categories, there is no conscious intention to deceive, this type of person would, if it suited his ends, purposely falsify a report. Similarly, if there is in the categories described above a wish to have a UFO report provide personal gain, it is a secondary and unconscious aspect of the experience. In the case of the psychopathic personality, however, a conscious wish for some sort of personal gain could very well be the primary motivation behind a false UFO report.

Apart from this type there are certain psychological phenomena which are not truly pathological in and of themselves but which occur under conditions that may be called altered states of consciousness, apperceptive deviances, or altered ego states. They are the hypnogogic phenomena which occur during the state of falling asleep, dreams, and hypnopompic phenomena of the awakening state. Also within this general category are eidetic images, *déjà vu* experiences, hypnotic or trancelike states, and the Isakower phenomenon described below. Withdrawal of the externally observing ego is common to all these introspective states, states when perception of the environment is highly diminished but ones in which environmental stimuli interplay with an intensified subjective awareness. An example of the interplay of the withdrawn ego and some aspect of its temporarily decathected environment occurs when the alarm goes off in the morning, and the external stimulus is incorporated into the dream content as the ego works to satisfy the wish to deny external reality and preserve sleep. Who among us has not had the experience of a dream so vivid that even after awakening we have difficulty establishing where the dreams ends and the reality begins?

Especially relevant to the subject at hand are the often vivid and amorphous mental images which occur at times when drowsiness and fatigue begin to overcome us and yet we remain awake. Descriptions include "whirling balls," "geometric forms," "black and fiery shapes" darting into the field of vision, or "whirling circles, flashing lights suddenly coming from everywhere." Similar to these hypnogogic phenomena are the vivid visual occurrences experienced when arising but not fully awake, during the hypnopompic state.

There are probably few among us who have not in a moment of spontaneous recall visually perceived a mental image, often a true memory, so vivid and so encompassing of attention that for the moment the environment dissipates in favor of the image. Such an "eidetic" image is indeed common. It is frequently reported by those undergoing psychoanalysis, particularly those who are in partially regressed states of consciousness. Another illustration of withdrawal of ego cathexis from the environment occurs during both self-induced and externally induced hypnotic states. Hypnotists state they can cause "the ego to withdraw into itself."

This brings us to the Isakower phenomenon. In 1938, Otto Isakower documented features of an unusual varient of the hypnogogic state.

> The visual impression is that of something shadowy and indefinite, generally felt to be "round" which comes nearer and nearer, swells to a gigantic size and threatens to crush the subject. It gradually becomes smaller and shrinks up to nothing. Sometimes there is fire somewhere.[3]

It was his contention, and is still a viable one, that this could be attributed to a recall of the subject's earliest infantile perceptions. More recent studies have demonstrated that the infant is intermittently perceptive and attentive from birth.[4] If the brain is the retentive computerlike organ that we feel it is, why should it not be able to recall any memory from any time, given favorable conditions? Certainly early, highly cathected, and repetitive experiences of the individual cannot be dismissed as simply lost forever. The favorable conditions for such a recollection exist during an altered ego state, one which allows the significant stimuli, recollections, and memory traces of the past to impinge themselves on consciousness. If we add to this the fact that sleep has a replenishing, a restitutive, in a sense a nourishing, function, might its onset not provide the favorable conditions for the emergence of gratifying, pleasurable, or reconstituting memories no matter how archaic? To this one can add the observations that it is easier to fall asleep with pleasant thoughts, and that overt sucking phenomena occur in many children and some adults as they drift off to sleep. It is for these reasons that Isakower and others suggest that this description is what the nursing infant experiences and that this recollection is of the breast as it presents itself to the infant.

Having allowed ourselves to come this far let us present the following descriptive account:

> The phenomenon was characterized in my experience by the typical amorphous mass, which, accompanied by moderate but ominous roaring, moved rapidly toward me from an immense distance. The mass was grayish or light tan in color. My attention was anxiously and inevitably forced upon its center which was either somewhat elevated, depressed, or whirl-shaped.[5]

Could this not be an account of a UFO experience? In fact it is a published description of an Isakower phenomenon personally experienced by a psychoanalyst. Isakower's original formulation was as follows:

> Yes, these imprints seem very easy to detect; they are the mental images of sucking at the mother's breast and of falling asleep there when satisfied. The large object which approaches probably represents the breast, with its promise of food. When satisfied, the infant loses interest in the breast, which appears smaller and smaller and finally vanishes away.[6]

The common denominator underlying all these phenomena is the withdrawal of ego from its attention to the external environment, with a subsequent though transient impairment of reality-testing. Under such conditions there can occur a regression, with the forgotten, the repressed, the unconscious coming to the fore. The perception experienced may seem to come from the environment, but in reality it is a projection from within. In the Isakower phenomenon, it is most clearly seen that the projection is in reality a symbolic representation for the breast.

Two of the major symbols of both the conscious and the unconscious world are the breast and the penis: "Thus the symbols of the breast vary accordingly from the vast lunar landscape . . . to the apples and pears of adult dream life." [7] The optical impressions of the earliest and highly cathected experience of life, the nursing situation, should and do provide a persistent set of imagery which is carried through life. Confirmation again is found in anthropology, sociology, and psychoanalysis. C. G. Jung has stressed that in "the round object—whether it be a disk or a sphere—we at once get an analogy with the symbol of totality well known to all students of depth psychology, namely the 'mandala' (Sanskrit for circle). This is not by any means a new invention, for it can be found in all epochs and in all places, always with the same meaning, and reappears time and again, independently of tradition." [8] While he was not speaking directly of the symbolism of the breast, the analogy holds just as it does for the moon-worshiping ancients. Similarly, more conventional forms of psychoanalytic thought confirm the importance of this imagery. Under conditions when regressive (primary-process) thinking or unconscious determinants of behavior seem to come to ascendancy, the possibilities for this type of imagery to become manifest are enhanced.

One can readily develop a series of more than circumstantial observations about the significance of the phallus:

> The history of phallic worship, which sprang from nothing more or less than normal exhibitionism in primitive man, demonstrates the mystic importance which was attached universally to the penis through the ages. Ever since antiquity, sex and especially the male sex organ has been deified more or less directly. In considering this subject, it is important to bear in mind that the notion of obscenity which later became connected with phallic cults was totally lacking at their inception.[9]

Thus, whether we approach the significance of the penis from its pleasure-giving capacities, or from its importance in the act of procreation, it is nevertheless, like the breast, a universal symbol, a highly cathected object, early on in the development of the individual, the culture, and the history of man.

The relevance of these observations about the breast and the penis becomes clearer when we look at "typical pictures" of UFO's. They are usually described as "saucer-shaped or cigar-shaped" objects (breastlike and phalluslike objects). Even more than providing a framework for understanding some UFO experiences, these considerations may also help to explain some of the emotionalism which surrounds the subject. The flying objects are embued with a psychological significance because they are representations, symbols, of highly libidinized primary objects in the development of the individual. They are symbols of extremes of gratification and of omnipotence. Culturally they are usually dealt with in suppressive and repressive ways.

Our intention has been to provide substance to a generally acknowledged but largely overlooked dimension in the evaluation and interpretation of UFO reports. Our observations have been limited to the mental processes of individuals, both those mentally healthy and those overtly disturbed. Studies suggest that a significant and perhaps increasing proportion of our population may be classified as having some diagnosable mental illness. Faced with high levels of environmental or intrapsychic stress, both groups, healthy and ill, may revert to more primitive modes of thinking, often characterized by magical explanations and symbolic usage.

A withdrawal of cathexis or attention from the environment, with faulty reality-testing, may occur in healthy people when falling asleep, or during other trancelike states. The Isakower phenomenon is a special instance of the latter. In the mentally ill, this withdrawal may express itself in the form of delusions or hallucinations. In either group the regression to more primitive modes of thinking allows the emergence of highly cathected images or symbols. Two such symbols are the penis and the breast, both of high phylogenetic and ontogenetic significance and related to concepts of omnipotence and omniscience. The fact that many UFO reports describe objects which are "cigar-shaped" or "saucer-shaped," penis- or breastlike, is suggestive that unconscious determinants may be of importance in some of these sightings.

In closing, we cannot avoid commenting on what appears to us to be an inordinate degree of affective heat generated among scientists involved in the study of the UFO phenomenon, whatever it is. None of us is ever as objective about our work as we often think we are. While intellectually we are acutely sensitive to the need for objectivity, we nevertheless to a greater or lesser degree narcissistically involve our own data, results, hypotheses, and theories. It follows, then, that to the extent that this is true, an attack on a man's work is affectively experienced by that man as an attack on himself. Thus, we expect in critical discussion of any scientific topic some bruising of feelings. What is extraordinary about the UFO problem is the degree to which feelings have become involved and polarized. Mature scientists accuse each other of publicity-seeking, of deceiving the public, of stealing documents, and in other ways of being dishonest, and they even threaten each other with lawsuits. It was not even possible to organize this symposium without arousing considerable passion. Clearly the affective involvement among people who are engaged in the study of this phenomenon is at an order of magnitude higher than it is in other kinds of investigation. This affect appears to be generated from the nuclear issue of whether or not at least some UFO sightings indicate the existence of extraterrestrial intelligent beings.

Psychoanalytic experience has demonstrated how often the anxiety generated by the same unconscious conflict will be handled by different people in ways that are diametrically opposed. For example, one person may successfully deal with anxiety arising from unconscious hostile-aggressive urges by becoming a sergeant in the Marine Corps, another, by devoting large amounts of time and energy to the pursuit of peace and other humanitarian goals. We suspect that the extraordinary affect generated by the UFO controversy derives from the fact that some common unconscious conflicts are being displaced onto it. Because again we have no clinical data at our disposal, we cannot know for certain what are the origins of the unconscious anxiety. However, if pressed, we would guess that it may derive from two areas of unconscious concern. We have already mentioned the possibility that repressed infantile sexual conflicts may play a part in some UFO experiences. Similarly for some scientists studying UFO phenomena, anxiety arising from these same unconscious conflicts may be dealt with, through sublimation and displacement, in terms of the issue of whether or not UFO's represent some strange form of life. It is also possible that some of the affective energy which is displaced onto the UFO controversy derives from the unconscious concern with death and immortality. To carry it a bit further into the realm of speculation, our guess is that for some of those who vehemently defend the extraterrestrial hypothesis it symbolically represents a denial of the finite nature of life. On the other hand, those who have a need to deny that there is any anxiety at all around the issues of death and immortality may be led to attack the hypothesis with considerable passion. While this extraordinary degree of involvement in these positions may have adaptive value for the individual partisans, it is clearly an obstacle to the effort to solve the UFO puzzle.

## NOTES

1. B. P. Dohrenwend and B. S. Dohrenwend, "The Problem of Validity in Field Studies of a Psychological Disorder," *Journal of Abnormal Psychology* 70 (Feb. 1965): 52–69.

2. P. B. Sheatsley and J. Feldman, "The Assassination of President Kennedy: Public Reactions," *Public Opinion Quarterly* 28 (1964): 189–215.

3. O. A. Isakower, "A Contribution to the Pathopsychology of Phenomenon Associated with Falling Asleep," *International Journal of Psychoanalysis* 19 (1938): 331–345.

4. H. Haynes, B. L. White, and R. Held, "Visual Accommodation in Human Infants," *Science* 148 (1965): 528.

5. G. Heilbrunn, "Fusion of the Isakower Phenomenon with the Dream Screen," *Psychoanalytic Quarterly* 22 (1953): 200.

6. Reference 3.

7. B. D. Lewin, "Reconsideration of the Dream Screen," *Psychoanalytic Quarterly* 22 (1953): 174–199.

8. C. G. Jung, *Flying Saucers: A Modern Myth of Things Seen in the Skies* (New York: Signet Books, 1969), p. 30.

9. N. K. Rickles, *Exhibitionism* (Philadelphia: Lippincott, 1957), p. 7.

motivation, and, in fact, in some cases, are frauds: the witness deliberately reported an incorrect observation for some personal gain. In other cases we will see that perception mechanisms, like eyes, have simply failed, through no fault of the witness; he was completely unaware of the fact that he made an incorrect observation. Of the four cases, two have to do with psychological motivation in witnesses and two have to do with perceptive processes.

First, we see in Figure 12-1 all the photographic records in existence of a UFO photographed by a thirteen-year-old boy who was out with a friend hiking through a cemetery one fall day. The event received no publicity. The boy had a Polaroid "Swinger" camera and took three photographs of the UFO. I have examined them and the "negatives," and the pictures have not been retouched. Can we make a judgment as to whether these pictures are valid evidence for a UFO? Or can we detect something which invalidates them and shows them to be fake? Of course, if the UFO is real, or if it is a fake and he has done the job well, the evidence won't be there and we will have to conclude that the results of our examination are inconclusive. I also interviewed the boy. The most important additional information I gained from that interview was that he reported that the size of the flying disk was two to three feet across, that it was a silver aluminum color all over, and that it made a whining noise, "slightly pulsating." (In fact, his mother reported that she heard this whining noise the following evening.) The boy said that the object moved about twenty miles an hour across a line of trees. He ran over and took three pictures of it; then it "went straight up." The Polaroid "Swinger" Model 20 camera has an exposure speed of 1/200 of a second and its fixed focus is good for objects from six feet to infinity. The first photo of Figure 12-1 shows the object as first seen (the white spot in front of the trees); then the boy got closer to take the second photo and finally got right under it just before it took off vertically in the lower photo. What does all this mean? Later I will give the answer.

The second event I want to discuss may be familiar because it is dredged up from the past in almost all new books on UFO's. It is the report of the reception in England of TV signals from the United States, three years after the signals had been sent. I first read about it in the December 1958 issue of the *Reader's Digest* (p. 186) which reported:

*Figure 12-1.* Three Polaroid photographs of a UFO, taken as described in the text.

> Have you ever wondered what ultimately becomes of the waves that radio and TV stations send out into space 24 hours a day? Do they fade and vanish or do they keep going forever? We do know that sometimes pictures appear mysteriously, long after a program has finished. One of the most famous of all such weird happenings was in England in September in 1953.*

The article went on to say that viewers in many parts of England saw the identification and call letters of TV Station KLEE in Houston, Texas (Channel 2) on their television screens—this was at a time when there were no satellites and transatlantic programing did not exist. The *Digest* article reported that "several viewers took pictures of the image to prove the happening." This last statement is true. What really startled the TV world, however—and this is the important part—was the fact that

* This statement is misleading; there has been only one such instance, the one described here.

when British broadcasting engineers advised KLEE in Houston of the unusual event,* they were told that the station had been off the air since 1950. No KLEE identification card had been shown for the past three years.

Where had that picture been for three years? Why did it appear only in England and how did it get back from wherever it had been? It does make you wonder, doesn't it?

KLEE had indeed broadcast from May 1949 to the middle of 1950, at which time the station was sold to another group. It continued to broadcast but with different call letters. It became KPRC and is still active as Channel 2 in Houston. KPRC maintains a great deal of correspondence on and records of this particular event and kindly made them available to me. I learned that the circumstances were investigated by engineers of the British Broadcasting Company, who said that they could neither prove nor disprove the validity of this particular event. For some reason engineers from the Chrysler Corporation in America also examined the event, and they pronounced it valid: the call sign of KLEE-Houston had actually been observed three years after its last transmittal. This led to the publication of the *Reader's Digest* article.

I learned more from the KPRC correspondence with the people in England who had seen "KLEE" on an English television screen. Also available was a large collection of photographs taken of the television screen. Of particular interest is a letter from one of the original observers addressed to the president of Station KLEE:

> Enclosed herewith is a photograph taken by an ordinary box camera of what I believe is your test signal received 3:50 p.m. 14 September 1953. It would be of great interest and help if you could be so kind as to confirm or deny by return mail that this is so and at the same time it would be of great help if you would endorse the back of the photograph and return. Your help in this matter would be much appreciated.

Subsequent investigation showed that it was a form letter and that letters like this had been sent not only to KLEE but to many other American television stations—each reported the observation of the call sign of the particular station.

The chief engineer of KPRC responded to this letter, although he did

* This part of the statement is untrue, as we shall see.

not know it was a form letter: "It is my belief that it definitely is not of the type test-signal that we have used even while we were KLEE. I believe that by chance this is a picture of a Kleenex advertisement. Kleenex is a facial tissue used by American women to remove makeup."

The alleged call-sign picture read "KLEE," and below the letters were two diagonal lines. The actual KLEE call sign contained the words "KLEE-TV" and underneath exhibited a map of Texas and the words "Channel 2, Houston, The Eyes of Texas." Obviously there are both differences and similarities. The letters "TV" are missing in the English picture. However, the diagonal lines are similar to the diagonal lines that radiate out of Houston in the real call sign. So one might argue (desperately) that the right picture was transmitted across the Atlantic, but with the wrong framing. Still, there was the Kleenex. KPRC sent a telegram to WCBS-TV, New York: "Please advise at your earliest convenience whether or not you ran a Kleenex commercial approximately 10:58 A.M., September 14, 1953." Actually this was logically absurd: for if indeed it was a Kleenex commercial, then the station received in England was not necessarily a CBS affiliate but any station from any network. There was a reply from WCBS, New York:

> In answer to your question, we did not run a Kleenex commercial at approximately 10:58 A.M., on September 14. We were carrying the Arthur Godfrey morning show and the sponsor for the 10:45 to 11:00 A.M. portion was Surf. It is possible that Godfrey might have mentioned Kleenex, but we have no records of it.

The Kleenex issue seemed to be destroyed as a working hypothesis, and so we return to the idea that observers truly had received KLEE in England.

This, of course, got KPRC even more excited. They wrote to the English observer, asking, "What kind of set are you using? Have you received any information other than call letters?" and many other questions. They received a lengthy reply from a gentleman who turned out to be a partner of the original observer in a business enterprise building television sets to receive messages from the United States. This was a great idea; in 1950 English television was terrible. The Englishman replied that he had taken the photograph and that he was using a set he had designed himself—a highly sensitive super-heterodyne set embodying the American tuning and an invention called a light cell; by rotating the cell away from the station, he got the best picture (a remarkable achievement!). The set had no antenna. The set, he said, gave poor images, but he had better photographs (this is a good trick!). He had no tape of the sound signal, because the sound was distorted by local interference.

In 1954, electronics engineers were convinced that this wonderful thing had, indeed, happened. In fact, KPRC has a large collection of the photographs of their call sign and all the other call signs received by this remarkable light cell super-heterodyne set. Looking at these, one is first amazed because they do indeed look like the call signs of American television stations. Then one wonders why there are only call signs; there is never anything else. Then one notices that they are very peculiar in some respects. There are usually white letters with some "snow," as one sees on real television, but on a perfectly black background. And in no case was there any snow in the black part. Second, every "A," and "P," or similar letter looks like a stencil letter. No picture has a "floating" black area in a letter. When one looks closer one detects that the noise on the letters looks very much like the grain in wood. In fact, the collection contains pictures with different call signs but with the same snow pattern. One begins to conclude that it was all a hoax, that a light was projected through stencils onto a piece of wood, and a photograph taken. The slides were then projected onto the screen of the impressive machine.

This conclusion became more certain when one read the correspondence in which the inventors claimed to pick up not only pictures from the United States but also from the Soviet Union, France, and South America. The call sign they had captured from Moscow television was written in English!

One might construct the following hypothesis: The Englishmen could have photographed call signs on television screens in various cities in the United States. Probably they lacked the skill to reproduce a moving picture on their receiver. Back in England, they saw that some of them were not very good, so they copied them with the stencil-projected-on-wood technique. They projected the slides on their machine before invited guests, and viewers, disbelieving, photographed the image on the machine and sent these photographs to the stations in the United States for verification; probably the form letter was supplied by the inventors. Perhaps this hoax had been carried on for years. Then one day, lo—disaster: A reply came saying, "That's the right picture but it hasn't been sent for three years." A station had changed its call sign.

At this point, everybody should have seen it was a fake. Instead they said this machine is even more wonderful than we ever believed! Some people are driven to make hoaxes, and many people, even highly educated ones, want to believe. They will accept any shoddy kind of evidence if it fulfills some dream.

To this day, the KLEE story is still reported as though it were a great mystery, despite the fact that the complete explanation has been in the possession of KPRC for years. People will work very hard to believe any kind of hoax, and there are people who will work to make such hoaxes.

Let's return to the UFO of Figure 12-1. No, it is not a Frisbee. Two things are suspect in these pictures, neither of which is conclusive. The first involves the fact that the object was described as moving twenty miles an hour and as being two feet in diameter. The camera shutter speed was $1/200$ of a second. In this time interval the object should move half an inch, and the object in the bottom photo should be blurred by half an inch, and it clearly isn't. But that is not conclusive because the witness said at one point it turned and went straight up when he was under it. Second, a careful inspection of the bottom two pictures reveals a similar forked branch of a tree apparently over the object in each case. But forked branches are not rare, so their appearance is not conclusive.

The witness had held to this story for many months. His parents believed him, his friends believed him; he was adamant that this had really happened. When I spoke to him he was a completely believable witness. There was not one statement or voice inflection that sounded phoney. He was bright, articulate, and completely credible. His story held together no matter how circuitously I came back to complex points. Nevertheless, there were the suspicions I have mentioned. The only one that could be tested was the tree-branch similarities. I asked the boy if he would go out with me the following Saturday and show me where he took the pictures. He agreed. That night the phone rang in my house, and on the phone was the boy in a very emotional state. It was all a fake, and he realized that when we got out in the field I would see that the same branch appeared in both pictures and, therefore that the object had not moved between the two pictures.

He had made a papier-mâché flying saucer in his basement, had tied it to the branch of the tree, and had gone out with his camera and had taken the pictures. He had built up the whole story and sold it to everyone. Even a young person can concoct an extremely good hoax.

In the next two cases, unusual phenomena actually occurred, and we were able to obtain nice calibrations of witness's observations, because we know exactly what happened. Extremely bright fireballs, brilliant meteors, burst in the skies in West Virginia about one month apart in 1962. Both metors appeared at about 10 P.M., so it was dark; yet a great many people were up. In each case there was a tremendous flash which lit up the whole sky almost like day. We also know that very loud sonic booms accompanied the fireballs but arrived many minutes later, so that the normal person might not associate the booms with the fireballs.

As a member of a group of astronomers from the National Radio Astronomy Observatory, I set out to find bits of the meteorites, if any had survived, and in the course of our search we interviewed as many witnesses as we could find. In the process we learned a great many things about witnesses, because we had, in effect, a controlled experiment.

The first fact we learned was that a witness's memory of such exotic events fades very quickly. After one day, about half of the reports are clearly erroneous; after two days, about three-quarters are clearly erroneous; after four days, only ten per cent are good; after five days, people report more imagination than truth. It became clear that later they were reconstructing in their imagination an event based on some dim memory of what happened. This is something that the UFO investigator rarely appreciates. The common procedure of starting a UFO investigation days after an event cannot lead to the most accurate description of the event.

Some aspects of perception are very accurate. For instance, the estimates of the duration of the fireball as it streaked across the sky were remarkedly accurate. In these cases it lasted four seconds, and the estimates were typically between three and five seconds, a remarkedly good performance. The estimates of the length of time until the sonic boom were also about right. This time it varied, of course, between one and five minutes, and the observers had the right order of magnitude: nobody said it was ten seconds; nobody said it was half an hour. The estimates were correct to within a factor of two.

One part of perception which clearly failed was the perception of the color of the objects. In this case we do not know for sure what color the objects were. Fireballs are normally greenish white, but they can vary in color depending on what materials are in them and the circumstances under which they are seen. In these cases, every color was reported, despite the fact that the objects were bright enough for color vision to work properly. The color reported most often was white, sometimes greenish white, but also orange, blue (the complementary color to orange), red, green, and "red-blue." Now remember that these events happened suddenly, as do many UFO events, to a dark-adapted eye. This may explain why some eyes saw different colors. In any case, when it comes to UFO's, the same errors will occur. The conclusion is that the eye, perhaps especially the dark-adapted eye, when presented with a bright unexpected light, may perceive any color. The colors reported are meaningless. This may, of course, explain why in many UFO reports there are conflicting color descriptions. Some people see red and green flashing lights while others may see yellow flashing lights at the same event; one should not discredit the observations because such conflicts could arise in a legitimate event in which everybody has seen the same thing.

Observers always overestimate their ability to establish the geometric position of an object in the sky; they are sure that they can recall where the object was and how fast it was moving. There is a tendency to overestimate the velocity. If witnesses had some reference points in their field of view, however, they can sometimes give an extremely accurate description, but the interviewer must use care in accepting this information. For example, we encountered many witnesses who had been in drive-in theaters when the fireball burst over the screen. They all thought they could give marvelous descriptions. We accompanied them to the drive-in theater, parked the car in the way it had been, and the observers sat back in the seat. Then the problems surfaced: they never really knew exactly where their car had been in the drive-in theater; changes in angle by as much as 45 degrees were possible. They had no good reference points connected with the automobile windshield. On the other hand, a game warden who was lying in wait for some deer spotlighters thought he could make a beautiful fix because he knew where he had parked his car and could remember just how he saw the fireball go through the trees. I went out with him and squinted through the branches and we got the fireball geometry completely wrong, yet he had been very clear about it. Then he remembered that at the time he had been eating his midnight snack and throwing sandwich and candy wrappers out the window, so we searched around and found these wrappers about a hundred feet from where he thought his car had been. We moved his car there and we got a beautiful fix. It was one of the most consistent.

Another couple who had been sitting on a porch swing in Covington, Virginia, were very good witnesses. They knew just how they had been located in relation to the furniture and the geometry was easily reproduced. The fireball seemed to cut across a corner of the porch, and so they gave a very good sighting. These experiences tell us that one must distrust any kind of geometrical information unless it is anchored or calibrated by some well-defined reference points.

The most curious thing we encountered was that a large percentage of the witnesses of both meteorites reported hearing a loud noise at the same time that they saw the fireball. Remarkably, the sound was always described as that of frying bacon, despite the fact that the witnesses had had no contact with one another. There was even one man who claimed —and this has been reported with other meteors—that he was inside the house and heard the sizzling sound; wondering what was going on, he went outside and saw the fireball. That is hard to believe because the event only lasted three or four seconds. One does not know how much credibility to give to such an account.

For information about one meteorite, which came eventually to be known as the "Mad Ann" meteorite, 78 witnesses were interviewed personally; of these, nine (12 per cent) reported hearing some sound simultaneously with seeing the meteorite. For the second, the Clarksburg meteorite, there were thirty-five witnesses interviewed (again, good statistics here); five (14 per cent) reported simultaneous sound. In fact, in the old meteor literature, with almost every fireball recorded, something like 14 per cent of the eyewitnesses report the simultaneous crackling sound, which should be physically impossible. How does the sound get there as fast as the light? The electromagnetic field strengths, from the flash of light, are not enough to generate a sound; they are no greater than with sunlight. Thus the phenomenon seems contrary to the laws of physics. Also, if you plot the locations of the people who hear these sounds you find that the timing of the sound does not obey the inverse square law, for a person at a great distance is just as likely to hear the simultaneous sound as one near the meteorite. One feels that this is a psychological phenomenon, a crossing of waves in the brain. Something like it is occasionally reported by drug users, although the fireball observers were almost certainly not drug users. I suspect under these unusual circumstances, with the mind not prepared for a stimulus but suddenly given an intense stimulus in one sense organ, there may be feed-through into other perceptive centers in the brain. Some will then not only see light but hear sound and perhaps even smell something. There is material here for an interesting experiment: to put people in a room and subject them to intense and abrupt light stimuli, perhaps in a form usually associated with a sound, and see what happens.

There are then at least two lessons to be learned from our investigations: One is that there is a need to carry out frauds and hoaxes—a desire to pull the wool over other people's eyes and to do it very cleverly for suprising reasons. The other is that even honest normal people make errors, because the human mind does not always have perfect sensors; it is an imperfect computer in dealing with the stimuli it receives.

Some aspects of perception work very well, some do well given certain qualifying conditions, and some fail completely.

# 13

## Influence of the Press and Other Mass Media

WALTER SULLIVAN

The "other mass media" in my title includes TV, exemplified by the *Star Trek* show, magazine articles and cartoons, comic strips, and paperbacks that overflow the bookstalls. There is no question but that these media, as well as the conventional press, have stimulated (some would say overstimulated) public imagination. In fact, the press invented the term "flying saucer" to describe what Kenneth Arnold saw in 1947. Many other such images have been similarly introduced—ghosts, for example, which entered folklore centuries ago. Few people in the United States believe in ghosts today, although children (and many grown-ups) enjoy a good ghost story. However, many American adults believe that UFO's are extraterrestrial visitors, and this makes a UFO report a good story; the public resonates to it.

Reporters earn their bread and butter with good stories, and don't get full credit if they "qualify to death" such a yarn. They are trained to check the source of an interesting report, then write it up "colorfully." But they don't do a full research job on it, and they hope that no one "shoots it down" before the readers can appreciate it. We journalists should not be too proud of this shallow treatment, knowing that deeper investigation will often lose the story. This even applies to science reporting at AAAS meetings when one speaker may have fascinating research results but other experts tend to qualify them and point out alternatives that, if reported, would simply confuse our lay readers. Of course, some rudimentary checking is essential to science reporting—as illustrated by the topic announced in an American Philosophical Society program some years ago. It read "The Moon as a Giant Electron," and seemed a real "hot" story until we established that "Moon" was a misprint for "Muon."

Constructive reporting of UFO stories presents a real challenge to the journalist. He must get to the scene in a hurry, apply a little sophistication, a little skepticism, a little human insight, and some scientific knowledge. More and more American newspapers now have trained science writers, and their improved scientific knowledge may account for the recent decline in press reports of UFO sightings; we screen out the obvious misconceptions. Another factor is that editors sense a shift of public interest, probably due to the successful Apollo program, and the general feeling, after following the astronauts by TV, press, and radio, that extraterrestrial visitors are highly unlikely.

Reporters have the opportunity to see at first hand the fallibility of human observations, and how a weak case can be made to look stronger by subtle wording. It is highly desirable to emphasize, as Dr. Hynek does (see Chapter 4), that we analyze what was reported, not necessarily what actually happened. Unfortunately, the UFO enthusiasts are not so careful; they speak of an object moving in an inexplicable manner, failing to mention that this is a report of a radar operator several years after the fact. Professional journalists, partly from fear of libel action, are trained to attribute such reports to a human source and to mention the circumstances which may cast some doubt upon them.

A significant point has been made by Dr. Hall (see Chapter 9), who said that many people "hold beliefs consistent with those of people around them," and that a strong system of belief "is likely to be defended vigorously beyond the point of logic." It is claimed that witnesses in many of the unexplained UFO cases did not believe in UFO's before their experience with them. My thesis is that we have all been conditioned by the press, radio, and TV—by the general tone of our society—to a hierarchy of beliefs that include for most of the population at least the image of UFO's.

One of my more humbling experiences illustrates this. After spending a year in Shanghai, I found, when I returned to the United States, that I had absorbed to some extent the point of view of that city's residents, for almost all of whom the Chinese Communists were the "good guys" and the Nationalists (who were bombing and blockading them) were the "bad guys." Another four years in West Berlin had a similar effect, and showed me that our American point of view is conditioned by subtle effects we are quite unaware of. Even in scientific research, there is the so-called observer effect whereby the results sought in an experiment are usually obtained, despite the investigator's precautions against bias. The classical example is Percival Lowell's observations of "canals" on Mars that have been thoroughly discredited by the Mariner close-up photos of that planet. Lowell's conscientious drawings of the canal network lent credence to the popular belief in intelligent life on Mars.

I can't match Dr. McDonald's careful investigations of UFO cases, but I have looked at the USAF Project Bluebook files, and found several interesting cases. One took place on March 3, 1968, when some seventy witnesses reported a UFO moving rapidly like a jet airplane. One woman in Indiana wrote that it was at tree-top level and "just a few yards away." "It was on fire at both ends" and had "many windows." At about the same time a woman in Ohio saw a similar object, which made her dog whimper and filled her with "an overpowering drive to sleep." Another woman in Tennessee saw square windows and riveted metallic fuselage on an object about 1,000 feet overhead, and an engineer driving on an Ohio highway was sure that the object changed course. All their reports refer to one object, identified by the USAF as the remains of the Soviet Zond IV booster which re-entered the atmosphere and streaked across the United States very like a meteor (much higher than tree-top level). Several pieces of debris may account for the reported windows. The reported behavior of the Ohio dog shows that humans often interpret animal reactions in a subjective way.

As a journalist at this symposium, I see a great deal of "noise" in the overall picture of UFO's. The enthusiasts focus attention on the rare cases which they claim support the hypothesis of extraterrestrial visitors. But they impose an enormous selection effect on the data. We must remember that billions of people look at the sky every night (and in the daytime, too). Although most of us in the indoor-living world only give the sky a glance, the farmers, sailors, astronomers, and night watchmen get more lengthy views. The vast majority of observations show nothing unusual, and the average scientist would ignore the few rare cases that deviate from the massive average.

We are also viewing the sky with electronic instruments more regularly today than ever before. I was associated with radar in its early days, and watched "bogies" do the most incredible things—changing speed and direction in ways we could not explain because radar tracking techniques in those days were far from precise. Dr. Hardy has described some of the more recent radar developments (see Chapter 7) and it would have been interesting to hear about the nationwide FAA system that monitors airlane traffic, or about the sophisticated radars used in antimissile defense, or in spacecraft tracking, to see whether they encounter cases like the UFO events in 1952 and 1957 (see pp. 90 and 56).

Public opinion has swung away from UFO's, and I think this helps account for the ridicule feared by witnesses who refrain from reporting their UFO sightings. It would be a disservice to science if everyone were conditioned to ignore strange sights in the sky. I can't believe that such sightings will be of world-shaking significance to science, but we should certainly not close our minds to the possibility of something interesting there. In all the scientific investigations I've studied, the scientist reaches conclusions that he considers the most probable interpretation of his data. Like Dr. Sagan (see Chapter 14), I consider the extraterrestrial hypothesis improbable, and prefer to look elsewhere. However, I disagree with Dr. Condon, who vehemently opposed this symposium, because I feel that UFO's represent a human phenomenon that is far more important than any of us realize. Our attitudes and perceptions are conditioned to a degree far beyond our capabilities of direct observation.

The advertising crowd know to some extent what can be done with conditioning, but what is the role of the mass media in fixing our political, ethical, artistic, or racial attitudes? What about the cold war and the gulf of mistrust that separates Moscow from Washington—or Moscow from Peking? The newspaperman contributes to the conditioning process—and hopefully to the understanding process—but he is just one part of the conditioning system. This system that forms men's attitudes and value judgments is an inchoate, involuntary thing that develops its own momentum. If the UFO's do nothing else, perhaps they will stimulate our academic friends to conduct meaningful sociological research on these matters. Perhaps they can save us from ourselves.

# PART IV

## RETROSPECTIVE AND PERSPECTIVE

# 14

## UFO's: The Extraterrestrial and Other Hypotheses

CARL SAGAN

There are many ideas which are charming if true, which would be fun to believe in, which are a delight to think about: reincarnation; the philosopher's stone to turn base metals into gold; the search for long or possibly indefinitely extended lifetimes; psychokinesis, the ability to move inanimate objects by thinking at them; precognition, the ability to foresee the future; telepathy, the ability to read somebody else's mind; time travel; leaving one's body (the literal meaning of ecstasy); becoming one with the universe. There is a wide range of concepts which would be fascinating especially if only they were true. But precisely because these ideas have charm, exactly because they are of deep emotional significance to us, they are the ideas we must examine most critically. We must consider them with the greatest skepticism, and examine in the greatest detail the evidence relevant to them. Where we have an emotional stake in an idea, we are most likely to deceive ourselves.

The idea of benign or hostile superbeings from other planets visiting the earth clearly belongs in such a list of emotion-rich ideas. There are two sorts of possible self-deceptions here: either accepting the idea of extraterrestrial visitation in the face of very meager evidence because we want it to be true; or rejecting such an idea out of hand, in the absence of sufficient evidence, because we don't want it to be true. Each of these extremes is a serious impediment to the study of UFO's; they affect different categories of people. A symposium such as this one must spend some time worrying about emotional predisposition.

I want to discuss first the extraterrestrial hypothesis of UFO origin, bearing in mind that its assessment depends upon a large number of factors about which we know little and a few about which we know literally nothing. What I want to lead up to is some crude numerical estimate of the probability that we are frequently visited by extraterrestrial beings.

There is a range of hypotheses which can be examined in such a way. Let me give a simple example: consider the Santa Claus hypothesis. This hypothesis maintains that, in a period of eight hours or so on December 24–25 of each year, an outsized elf visits fifty million homes in the United States. This is an interesting and widely discussed hypothesis. Some strong emotions ride on it, and it is argued that at least it does no harm. We can do some calculations. For example, suppose that the elf in question spends one second per house. This isn't quite the usual picture—"Ho Ho Ho" and so on—but I imagine he is terribly efficient, and very speedy; that would explain why nobody ever sees him very much. With $10^8$ houses he has to spend three years just filling stockings. I've assumed he spends no time at all in going from house to house. Even with hyper-relativistic reindeer, the time spent in $10^8$ houses is three years and not eight hours. This is an example of hypothesis testing independent of reindeer propulsion mechanisms or debates on the origins of elves. We examine the hypothesis itself, making very straightforward assumptions, and derive a result inconsistent with the hypothesis by many orders of magnitude. We would then suggest that the hypothesis is untenable.

We can make a similar examination, but with greater uncertainty, of the extraterrestrial hypothesis which holds that a wide range of unidentified flying objects viewed on the planet Earth are space vehicles from planets of other stars. The report rates, at least in recent years, have been several per day at the very least, but I will make the much more conservative assumption that one such report per year corresponds to a true interstellar visitation. Let's see what this implies. To pursue this subject we have to have some feeling for the number, *N*, of extant technical civilizations in the galaxy—that is, civilizations vastly in advance of our own, civilizations which are able by whatever means to perform interstellar space flight (I will say a word about the means later, but the means don't enter into this discussion just as reindeer propulsion mechanisms don't affect our discussion of the Santa Claus hypothesis).

An attempt has been made to specify explicitly the factors which enter into a determination of the number of such technical civilizations in the galaxy. I will not here run through what numbers have been assigned to the various quantities involved—it's a multiplication of many probabilities, and the likelihood that we can make a good judgment decreases as we proceed down this list. *N* depends, first, on the mean rate at which stars are formed in the galaxy, a number which is known reasonably well. It depends on the number of stars which have planets, which is less well known but there are some data on that. It depends on the fraction of such planets which are so suitably located with respect to their star that the environment is feasible for the origin of life. It depends on the fraction of such otherwise feasible planets on which the origin of life in fact occurs. It depends on the fraction of those planets on which the origin of life occurs in which, after life has arisen, an intelligent form comes into being. It depends on the fraction of those planets in which intelligent forms have arisen which evolve a technical civilization substantially in advance of our own. And it depends on the lifetime of the technical civilization. It's clear that we are rapidly running out of examples as we go further and further along. That is, we have many stars, but only one instance of the origin of life, and only a very limited number—some would only say one—of instances of the evolution of intelligent beings and technical civilizations on this planet. And we have no cases whatever to make a judgment on the mean lifetime of a technical civilization. Nevertheless there is an entertainment (which is the way I put it) which some of us have been engaged in, making our best estimates about these numbers, and coming out with a value of *N*. The equation which comes out [1] is that *N* roughly equals 1/10 the average lifetime of a technical civilization in years. If we put in a number like $10^7$ years for the average lifetime of advanced technical civilizations, we come out with a number for such technical civilizations in the galaxy of about a million: that is, a million other stars with planets on which today there are such advanced civilizations. Now I think you will recognize that this is quite a difficult calculation to do accurately and moreover that the choice of $10^7$ years for the lifetime of a technical civilization is rather optimistic. But let's take these optimistic numbers and see where they lead us.

Let's assume that each of these million technical civilizations launch $Q$ interstellar space vehicles a year; thus, $10^6Q$ interstellar space vehicles are launched per year. Let's assume that there's only one contact made per journey. In the steady-state situation there are something like $10^6Q$ arrivals somewhere or other per year. Now there surely are something like $10^{10}$ interesting places in the galaxy to visit (we have several times $10^{11}$ stars) and therefore at least $10^{-4}Q$ arrivals at a given interesting place, let's say a planet, per year. So if only one UFO is to visit the earth each year, we can calculate what mean launch rate is required at each of these million worlds. The number turns out to be 10,000 launches per year per civilization; and $10^{10}$ launches in the galaxy per year. This seems excessive. Even if we imagine a civilization very much further advanced than ourselves (I'll mention in a minute that it's a considerable undertaking to travel effortlessly between the stars), to launch 10,000 such vehicles for only one to appear here is probably asking too much. And if we were more pessimistic on the lifetime of advanced civilizations we would require a proportionately larger launch rate. But as the lifetime decreases, the probability that a civilization would develop interstellar flight very likely decreases as well.

There is a related point made by Hong-Yee Chiu; [2] he begins with more than one UFO arriving at Earth per year, but his argument follows the same lines as the one I have just presented. He calculates the total mass of metals involved in all of these space vehicles during the history of the galaxy. The vehicle has to be of some size—it should be bigger than the Apollo capsule, let's say—and you can calculate how much metal is required. It turns out that the total mass of half a million stars has to be processed and all their metals extracted. Or if we extend the argument and assume that only the outer few hundred miles or so of stars like the Sun can be mined by advanced technologies (further in it's too hot) we find that $2 \times 10^9$ such stars must be processed, or about 1 per cent of the stars in the galaxy. This also sounds unlikely. Now you may say, "Well, that's a very parochial approach; maybe they have plastic spaceships." Yes, I suppose that's possible. But the plastic has to come from somewhere, and calculating plastics instead of metals changes the conclusions very little. This calculation gives some feeling for the magnitude of the task when we are asked to believe that there are routine and frequent interstellar visits to our planet.

Let me say a few words about possible counterarguments. For example, it might be argued that we are the object of special attention: we have just developed all sorts of signs of civilization and high intelligence like nuclear weapons, and maybe, therefore, we are of particular interest to interstellar anthropologists. Perhaps. But we have only signaled the presence of our technical civilization in the last few decades. The news can be only some tens of light years from us. Also, all the anthropologists in the world do not converge on the Andaman Islands because the fishnet has just been invented there. There are a few fishnet specialists and a few Andaman specialists; and these guys say, "Well, there's something terrific going on in the Andaman Islands; I've got to spend a year there right away because if I don't go now, I'll miss out." But the pottery experts and the specialists in Australian aborigines don't pack up their bags for the Indian Ocean.

To imagine that there is something absolutely fantastic, you see, about what is happening right here goes exactly against the idea that there are lots of civilizations around. Because if there are lots of them around, then the development of our sort of civilization must be pretty common. And if we're not pretty common then there aren't going to be many civilizations advanced enough to send visitors.

There is another argument: namely, that the space vehicles that are allegedly being seen are in fact just the local craft—the shuttles that come from some large mother ship which is the real interstellar space vehicle. (Drs. Grinspoon and Persky may be interested to hear that the vehicles in the UFO literature described as "mother ships" are the ones that are cigar-shaped, and I shudder to think what that means for their interpretation.) But again the mother-ship idea changes things by factors of 10 or 100 at the very most; it doesn't resolve our problems.

So I deduce from these arguments that the extraterrestrial hypothesis is in some trouble if we're to imagine that even a smallish fraction of the ten or twenty thousand UFO cases reported in the last twenty to twenty-five years are interstellar in origin.

So far, I've said not a word about the methods of interstellar transport. There are serious problems in interstellar flight, principally because the space between the stars is enormous. There are a large number of stars—about two hundred billion stars in our galaxy alone. There are at least a million other such galaxies. But the average distances between stars in our galaxy is a few light years; light, faster than which nothing that can slow down can travel, takes years to traverse the distances between the nearest stars. Space vehicles take that long at the very least. In order for a space vehicle to get from one star to another in a convenient period of time it has to go very fast. It has to go very close to the speed of light so that relativistic time dilation can enter into the problem, and so the shipboard clock can run more slowly compared to a clock left on the launch planet. To travel very close to the speed of light is difficult. There is a literature on the subject of relativistic interstellar flight, maybe thirty or forty papers in various scientific journals.[3] It is easy to see that carrying sufficient fuel for an interstellar flight is really out of the question, even if the fuel is half matter and half antimatter (never mind what's holding the antimatter). The ratio of mass of fuel to mass of usable payload that is required in such ventures is prohibitively large.[4] An alternative has been suggested by Bussard:[5] an interstellar ramjet with enormous frontal loading area which collects interstellar material on the way, accelerates it out the back, and therefore does not have to carry its own fuel. It doesn't run into the mass-ratio problem, but it does run into some other problems. The point of the Bussard ramjet is not that it is practical, for it surely isn't that: building a spacecraft several hundred kilometers across is only an engineering problem but it's not an engineering problem that's going to be solved tomorrow. But the Bussard ramjet does overcome this mass-ratio difficulty, which involves fundamental physics. There have been some recent discussions of Bussard's idea, for example one by Fishback[6] which critically assesses the ramjet concept and judges that there are various mechanisms including magnetobrehmstrahlung and problems in the turbulence of the stream that comes out the ramjet that makes stability at high velocities very difficult. But this is a second-order criticism. What I've learned from the Bussard idea is that it is possible even at the present time to think of methods of running between the stars. The fact that none of them may work well is, I think, not critical. What is critical is that there are conceivable ways of doing it without bumping into fundamental physical constraints. And this suggests that it is premature to say that interstellar space flight is out of the question.

And so now I turn in the other direction. I believe the numbers work out in such a way that UFO's as interstellar vehicles is extremely unlikely, but I think it is an equally bad mistake to say that interstellar space flight is impossible. One may argue that space flight is not the most cost-effective way to communicate between civilizations and that interstellar radio contact is a better way. Or one can imagine a wide range of other possibilities—neutrino transmission, modulated gravity waves, tachyons; next year we'll think of some more. But such considerations do not exclude interstellar flight. We do not know enough to exclude such visitations, but the probability of such visitations seems very small.

If all this is true (and even if we were to admit the possible existence of very strange and very reliably reported cases), why is it that the extraterrestrial hypothesis of UFO's is so popular? Why is it even around? There is a wide range of other perhaps equally plausible hypotheses that we don't often hear about. Why is there no faction that urges that an unidentified flying object is a projection of mankind's collective unconscious? Psychiatrists have written on the collective unconscious; why not that? Or time travelers? Or visitors from another dimension? Or the halos of angels? Or apparitions from the spirit world, or from Middle Earth, or Witchland, or Perelandra? There are a wide range of possibilities that could be thought of. How about harbingers of divine wrath? If only we could interpret them properly! Or fulfillments of prophecies from the Bhagavad Gita? My question is: How can *these* possibilities be disproved? What is the critical test for disproving the hypothesis that UFO's are angels' halos? It's difficult to think of a really critical test.

It seems to me that one runs into precisely the same problem with the extraterrestrial hypothesis. There is no good empirical test which could, for all cases, exclude this hypothesis. I would like therefore to ask: is it possible that we hear so much of this hypothesis because the idea of extraterrestrial visitation somehow resonates with the spirit of the times in which we live?

There are four such resonances that I can think of: religious aspects, the relief of boredom by believable novelty, military classification, and intolerance of ambiguity. I think it is pretty clear that over the last few centuries science has systematically expropriated areas which are the traditional concern of religion. It used to be that the opening of the flowers was due to the direct invervention of the Deity. That is not a view now often heard. As Darwinian evolutionary views became popular and mechanistic interpretations of the origin of the solar system and of cosmology became widely disseminated, part of the traditional domain of religion contracted, whether for good or for ill. At the same time, traditional forms of religion have been a very firm portion of nearly every culture of mankind; it is unlikely that the needs for belief in the gods, whether valid or not, can be destroyed so easily. In a scientific age what is a more reasonable and acceptable disguise for the classic religious mythos than the idea that we are being visited by messengers of a powerful, wise, and benign advanced civilization?

I have some direct experience with a few UFO cases in which this sort of thing is clearly part of the motivation for both exaggerating and denigrating the sightings. I certainly don't maintain that unfilled religious needs are responsible for all typical UFO sightings, but such needs are a likely resonance between the extraterrestrial hypothesis and quite unscientific aspects of the problem. Incidentally, I believe this view is politically dangerous: The expectation that we are going to be saved from ourselves by some miraculous interstellar intervention works against the necessity for us to solve our own problems.

The second point is the question of boredom and novelty, which I can best illustrate with a brief personal experience. Once when I was on the faculty at Harvard I gave a popular lecture on something or other, and in the question period at the end there were some questions about UFO's. I said that I felt at least a great fraction of them were misapprehended natural phenomena. For some reason that I don't understand, policemen are present at all such public gatherings, and as I walked out after the last question, two policemen outside the lecture hall were pointing up at the sky. I looked up and observed a strange brilliant light moving slowly overhead. Of course, I got out of there fast, before the crowd came out to ask me what it was. I joined some friends at a restaurant and said, "There's something terrific outside." Everyone went outside. They really liked it—it was great fun. There it was. It wasn't going away. It was clearly visible, slowly moving, fading and brightening, no sound attached to it. Well, I went home, got my binoculars, and returned. Through the binoculars I was able to resolve the lights; the bright white light was really two closely spaced lights, and there were two lights on either side, blinking. When the thing got brighter we could hear a mild drone; when the thing got dimmer, we couldn't hear a thing. In fact it turned out to be a NASA weather airplane. When I showed my friends at the restaurant that what we were seeing was in fact an airplane, the uniform response was disappointment. I mean, it's no fun to go home and say, "You'll never guess what happened. I was in this restaurant, there was a bright light outside, it was an airplane." That's not a memorable story. But suppose no one had a pair of binoculars. Then the story goes, "There was this great light out there and it was circling the city and we don't know anything about it. Maybe it's visitors from somewhere else." That's a story worth talking about. Despite all the novelties of our times, there is a kind of drudgery to everyday life that cries out for profound novelties; and the idea of extraterrestrial visitation is a culturally acceptable novelty.

The third point is classification. There is in our society, I think everyone will recognize, a certain paranoid aspect to some of the UFO cases. (The paranoia in our society is not, of course, restricted to UFO cases.) For example, there was a feeling in some circles a few years ago that NASA was keeping to itself photographic evidence which showed that the earth was not round. There were buttons one could buy in Berkeley which said something like "Where are the photos of the whole earth?" There were many space vehicles that went up and took pictures of the earth but they were always close-up pictures—continents, oceans, but never the entire spherical earth. Well, when space vehicles flew far enough from the earth to take such pictures, they were taken—and sure enough, the earth was round.

In such a climate of opinion Air Force classification of UFO reports resonates exactly. The armed forces have a tendency to classify everything in sight, including, or maybe even especially, bizarre cases which are inadequately examined and which involve military personnel. Then the fact that such cases are classified starts rumors. Somebody who is in a position to know realizes the Air Force does have relevant data; and it is just a short step to the idea of official conspiracy to suppress the truth. Had the data not been classified, then independent scientific judgments would have been possible. In many cases, such independent scientific analysis would show that the cases have a natural explanation. The culprit is classification. I have a friend who says that in America today if you're not a little paranoid you're out of your mind. The military has a responsibility not to add further to the paranoia.

The fourth point is a widespread intolerance for ambiguity. It's more difficult to keep two ideas in my mind than it is to keep just one idea in my mind. This point comes out very clearly if you've ever written a popular book on science. I did that once for a major publishing enterprise. I would write, "Here's the observation; some people think this is the explanation, some people think that is the explanation." The story I would get from their editors would be "Don't bother me with the alternatives; just tell me what's true." I think it's a fact of life that many people are uncomfortable with ambiguity, with a judgment withheld. But, it seems to me, this is precisely where we ought to be on the UFO problem: to say that there aren't enough data, that good judgment isn't possible yet, and that an open mind should be kept. Scientists are particularly bound to keep open minds; this is the lifeblood of science.

As a concluding word let me say that I believe the search for extraterrestrial intelligence to be an exceedingly important one both for science and for society. It is difficult to think of a more important scientific question. But I do not believe that the most efficient method of examining this topic is via the UFO problem. The best hope for such investigations is NASA's unmanned planetary program and attempts at interstellar radio communication.

## NOTES

1. I. S. Shklovskii and C. Sagan, *Intelligent Life in the Universe* (San Francisco: Holden-Day, 1966; New York: Delta Books, 1967).

2. H.-Y. Chiu, book review, "The Condon Report, Scientific Study of Unidentified Flying Objects," *Icarus* 11 (1970): 447.

3. C. Sagan, *Planetary and Space Science* 11 (1963): 485, reprinted in part in *The Coming of the Space Age,* ed. Arthur C. Clarke (New York: Meredith Press, 1967).

4. E. Purcell, *Brookhaven Lecture Series* 1 (1960): 1; S. von Hoerner, "The General Limits of Space Travel," *Science* 137 (1962): 18.

5. R. W. Bussard, "Galactic Matter and Interstellar Flight," *Astronautica Acta* 6 (1960): 179.

6. J. Fishback, "Relativistic Interstellar Spaceflight," *Astronautica Acta* 15 (No. 1) (1969): 25.

# 15

## The Nature of Scientific Evidence: A Summary

PHILIP MORRISON

It is most difficult to give an adequate summary of this data-full and detailed discussion which has occupied three sessions. Therefore, I will not try to make a detailed point-by-point synopsis. My original task was to prepare a commentary on method, into which I have tried to introduce the data and the issues argued before us at this symposium, to present a kind of model of what I think can be said. I want to stress that this is not a final position. Like all, or nearly all, scientific positions, it is tentative and ambiguous in nature; for that reason it will fall short of conviction, as indeed it should.

I discovered with some alarm, when I read the printed version of the AAAS program, that the title of my contribution was the "Nature of Physical Evidence." What I meant to say was "The Nature of Scientific Evidence." As a physicist, I put science in the domain of physics; I did *not* mean to distinguish physical evidence from testimony. I admit that distinction is made in the common interpretation of the phrase, especially in common law. But this is a fortunate error, because it enables me to view the topic more clearly than I had earlier.

What is the nature of scientific physical evidence? There are many glib canons given in the books. For example, reproducibility is often listed, especially in elementary accounts. This is absurd; no one wants to be asked to reproduce an eclipse or an aurora or very many other natural phenomena. It would wipe out many sciences on the earth if this canon were taken seriously. The rule doesn't really mean reproducibility in the sense of a laboratory-replicated experiment. That of course is a very neat model, but we can't always use it.

Nor does it mean—as I think Dr. Baker, in his analysis of photographs, was justified in saying (I think he would quite agree that this is not only a philosophical condition but a working one for him)—that scientific information, scientific evidence, must be quite "hard," in the sense that it must be recorded objectively, without the specific intervention of a witness. Where would Darwin have been were this canon accepted altogether? In fact, a great many observations are not of this reproducible, instrumental kind. Indeed, it is probably peculiar to our own time, when we have powerful sensors and powerful ways of handling data in electronic circuitry, to think of things in this way. When Eddington was trying to characterize the same thing for the popular reader, he spoke of pointer readings; and of course the witness's view of the coincidence of the pointer with a scale marking was a necessary link as Eddington saw it. Photography or print-out was not then commonplace in recording meter readings.

So neither reproducibility nor the absence of humans in the data link is the criterion for good evidence. But I submit that there are such criteria, and I would like to elaborate on them. Perhaps it is not desirable to distinguish between excellent evidence in science and excellent evidence in the law. I don't wish to make that distinction very sharply: they may turn out to be the same thing. But in the sciences, at least we can carry out a detailed and self-conscious analysis of the competence of the instrument to support the inferences drawn. The hardest point for students to realize (especially when they have had a long history of schooling in excellent textbooks, in which theorem after theorem is stated, and inference after inference drawn, and all experiments are nicely described) is that, in fact, the experiments and inferences printed in scientific literature are by and large false! I am not talking about reports of UFO's but about the sort of detailed dull material that appears on page after page of our kind of journals. Usually there are homely thoughts, usually the authors have some hint of the fact that somewhere mixed in something is wrong. Now, what one has to understand to use an instrument reliably is the full chain of events that causes the needle to occupy the scale-position—or that causes a "print-out" to occur, to use a more modern idiom. I submit that the effective personal witness needs to be examined in exactly the same way.

From the point of view of drawing inferences about events, a witness is simply an extraordinarily subtle and complex instrument of observation. I think the scientist has no other way of looking at his witness's testimony than that. From what is ordinarily a statement, verbally or in writing, possibly with gestures and emotional tone, we have to infer those perceptual and other causes in the history of the individual which are competent to produce at the end of this very complex chain the statement that results. That is treating the witness instrumentally, and if we wish to derive information we can do no less. Those who have reported on witnesses repeatedly found themselves, quite accurately saying that the witness reported the object was one or two miles away, or some such phrase. Of course the first question is: "How could the witness have judged, given a broad knowledge of the input channels that he normally has available, how distant the object was?" Anyone who has tried to elicit this, especially from persons wholly unschooled in estimations of this sort, or indeed in mathematical calculations of any sort beyond commercial transactions, knows that the idea of an angular ratio is hard for people to grasp. It is a subtle notion; I am not at all sure that it was clear in the mind of Aristotle, who was a man of extraordinary ability. I think a witness's statement should be regarded in much the same light as the reading of a barometer or the print-out of a computer: a large number of judgments, inferences, assumptions, and hypotheses are necessary to interpret it. The analysis of that chain is the essential feature of scientific evidence. Without that discussion, there is no scientific evidence; nor can a sound conjecture or even a suggestive hypothesis often be made. The evidence comes from asking, "Is our present inference consistently drawn from what we admittedly know?"

I find it most interesting to study, from this point of view, what was the decisive paper in a long chain of skeptical attitudes held by science over a phenomenon which is now undoubted. That paper is much cited but rarely read, judging from the fact that the original print, in the Widener Library stacks at Harvard, has not circulated for fifty years, as far as I can tell: a little brochure 170 years old; an engaging, beautifully written, and deeply argued paper by Jean-Baptiste Biot published in the year 11 of the Revolution and beginning, "On the 26th of Florial, I left Paris for the West." [1] It gives his account of the voyage he took to the Normandy village in the year we call 1803, on the date we call April 26—all renamed in the enthusiasms of the Revolution—to study the report that indeed stones had fallen from the sky on that occasion. He was a little slow; he came three weeks or so after the event, the best he could do; he was 120 miles away and the news was slow to travel.

Biot was one of the most able natural scientists in Paris. His work encompasses electromagnetic theory, Egyptology, astronomy of China, calculational astronomy, orbit calculations—an enormous bibliography of sound publications. He had a few days' notice, so he prepared himself by talking to mineralogists in Paris and by examining the collections of the museums, especially of those objects long claimed to be thunderbolts, stones that fell from the sky, and so on. These are admirable preparations, if slight on our scale. He went to L'Aigle, and wrote the report in a charming personal style. I'll try very briefly to summarize the kind of evidence that he found adequate and indeed which the scientific world found adequate, because after Jean-Baptiste Biot's report no serious doubt remained in the scientific literature. (Several years before that time, though, Chladni had in a large tome made many of the same points, but his work was not generally accepted.)

Biot heard the following tale: People had seen a daytime fireball which left a persistent trail, heard several loud detonations, found marks on the ground and on trees, and collected debris. And he went to see for himself what these reports amounted to. Traveling around the four or five villages in the five-mile by one-mile area from which these reports emanated, from which he could collect samples, he acquired or saw two to three thousand separate pieces of rock. All of these rocks were of a similar kind and texture to his eye and magnifying hand lens. They were of a kind never seen in that region before, and they had not been reported on the geological survey of that region that had by chance been performed only five or six years before. There are no volcanoes in the region, and there were no piles of slag or furnaces or glassworks producing a kind of steady outflow of recently melted or strangely textured stony or ceramiclike material. The material contained volatiles and had an odor. It changed in appearance and hardness with time, as he verified by observing samples for a few weeks. From these things he concluded, I think unquestionably, that there had been recently delivered to the villages in Orne a strange rock of recent vintage which had arrived in a new way. It was like several, but not all, of the samples held in Paris which were claimed to be thunderbolts—a fact of which none of the persons in the villages had had any knowledge. Where would they have gotten such rocks? So he felt it was unmistakable that the new rock was delivered somehow in a curious way.

Second, he considered the witnesses. He counted the witnesses and spoke to a large sample of them. I can't establish that he spoke to every one of them, but he says there were hundreds of eyewitnesses of the fireball, trail, detonation, marks, and debris. They spanned all social orders: the mayor, priests, soldiers, workers in the fields. All, as he said, independent. Many of them did not know each other and had not spoken to each other; there was much agreement on the events they had seen—not complete agreement, but general agreement on time and place. And there was no interest he could see on the part of these persons to deceive him; he was sensitive to that. Finally, he found for himself evidence of fires, broken branches, even some broken portions of house roofs which no one had noticed before and which he could interpret as caused by this fall. Such was the evidence on which he based his conclusion. I think the canons he implicitly followed are the ones that I recommend. He looked for independent and multiple chains of evidence, each capable of satisfying a link-by-link test of meaning. That is the *sine qua non* of responsible evidence. If we are to believe any hypothesis, however plausible or implausible, concerning new events—particularly those that do not satisfy the easy quality of being reproducible at will by those who undertake to set up a laboratory for the purpose—then we must find a case as clearly filled with multiple, independent chains of evidence satisfying a link-by-link test as Biot's case in Orne.

The quantitative canons for acceptance of a new phenomenon will vary, of course, but I submit that if we remain in the presence of ambiguity for a long time in these matters, that is as it must be. Only a fortunately clear example, I think, can demonstrate the hypothesis. The story of the acceptance of erroneous hypotheses is long. The story of excellent discoveries which were at first disbelieved but which we now can prove is also a long story, too long to be handled here.

I should like now, in this light, to mention a few of the kinds of examples we've heard here. Most of these are accounts I myself have been told in the two years or so that I've been sensitive to the problem of UFO's. First, there is the question of testing link-by-link the nature of strong evidence, and it is extremely hard to do this retrospectively. This is the hardest problem, I think, that investigators like Professor Hynek and Dr. McDonald face. The best they can do is to work a cold trail. It was also the best Biot could do. His trail was warmer than usual, and it was complicated; the footprint of that fireball was very strong in the villages.

The best example I know of an extraordinary UFO event—hard evidence objectively recorded on photographic film—was displayed on television programs several years ago. I find it quite a remarkable example. It was accompanied by a statement by Charles Gibb-Smith of the Institute of Aeronautics Museum in Kensington in London, who is a scholarly and precise historian of aviation, familiar with construction of aircraft over the entire span of aeroengineering. After having seen the motion picture film, he submitted that it was a picture of a metallic object, that it was of some sort of aircraft, that it resembled no aircraft that he knew of in the history of aeronautical engineering, and that its rate of recession exceeded anything that anyone had seen before. He drew this conclusion on the basis of the 16 mm film which was shown on television, so that we all could see the same evidence. That evidence was quite plain. There were many good arguments, which I shall not give, which made most people feel that the film was in fact not a hoax, not deliberately produced to deceive. What the photograph showed was a view from the inside of a passenger aircraft, looking out. Outside the window was a tiny metallic ellipsoid, whose major and minor axes increased rapidly together until they reached a considerable subtended angle, when the texture and illumination resembled very strongly the metal constructions of an airplane; then rapidly the image disappeared, the whole thing occupying some 0.1 of the field of view at its maximum, and in a few seconds going from invisible to 0.1 the field of view and back again. A very striking observation indeed, and certified by a legitimate expert not to be any kind of aircraft anyone has ever seen!

Well, of course, he was almost right about that. What he saw was in fact the tail of the very aircraft in which the camera was riding, perceived through the extraordinarily astigmatic lens of the thick edge of one of those round plastic windows set into the pressure cabins in some aircraft. The image was not merely distorted but topologically distorted. As the aircraft tilted, a piece of the tail structure came in view of the lens, rapidly grew, and rapidly went out of view again. It *was* a metallic textured object, and it was flying.

Now there the argument was very plain. What you had to admit was, "Yes, there is an image; it's on the film." Let us waive the question of whether it's legitimately there or was assembled through several events. Assume that it's present at one time. How did the image reach the film? The film was exposed through the camera lens, and we understand the competence of camera lenses. But the camera lens was separated from the outside world by the window of the aircraft through which the light passed. Thus the next question is: "Could the window of the aircraft have had anything to do with it?" And the answer, of course, is that the distorting edge of the window can produce very strangely modulated images. That is an example of what I call the link-by-link test of one striking chain of evidence.

It's very difficult to establish the credibility of a witness. In law we find that when any kind of technical or substantive information must be presented, there is normally on each side of the case a sworn witness, of high standing in the community, of tested competence in his field, who often disagrees with the judgment of another witness with similar qualifications. Since in general (not always) there is only one answer to such a question, I submit, on the logic of the matter, that one or the other of these people—creditable, skilled, and competent—is wrong, in spite of this link-by-link test. Here I will disagree, for example, with Dr. Hynek, who feels he can establish credibility for a single witness. I would say that *no* witness is credible who bears a sufficiently strange story. The only hope is for independent chains, several independent witnesses, and then the credibility certainly rises. Moreover, independence is most important (I shall return to this point). I want to emphasize that the singleness of a witness necessarily puts his case into some sort of doubt. All of us know how people have been mistaken with the best will in the world.

First, I want to introduce what is the weakest argument of all (I will indicate them in order of decreasing weakness): the question of what is classically called sufficient reason; namely, you do not simply multiply hypotheses, you try to get by with the least you can. This is a purely economic criterion; it is not the only guide to science—it is by no means a safe guide to science. It is merely one guide, but I find it quite interesting and I think the following anecdote will illuminate what I mean:

There is a whole body of literature—not the UFO literature but in a literature very close to it—concerned with the strange disappearance of ships. The number of ships that were well sailed and well built, that encountered no bad weather or other obvious explanation, but that nevertheless mysteriously disappeared, is large. Many disappeared in certain parts of the world, systematically, where there is heavy traffic. Many hypotheses can be evolved to explain these disappearances. The most popular one when I was a boy, and enormously impressed with this literature, was that there was a class of beings who effectively swallowed ships, who came and got the ships and took them home for samples. It occurred to me at once that this was a perfectly plausible hypothesis, and I could not on *a priori* grounds exclude it. Possibly some of the arguments that Dr. Sagan made (Chapter 14) might have done so, in terms of time and ability. I'm pretty wary about excluding one disappearance every few years or decade on such grounds. But it is interesting to ask a related question: "How many trains have disappeared? How many buses, coaches, wagons, and so on?" Well, it turns out to be very few by comparison; indeed, it's hard to find any at all. And when you ask what distinguishes trains from steamships, the answer is very plain. When something strange goes wrong enough to sink a ship, there is almost no way to know what happened, because the ship is 5,000 or 15,000 feet below the surface. If something goes wrong with a train, there it is all over the New York Central right-of-way. Somebody has to cope with it, and they nearly always find out what happened.

I told this generalization to an extremely able colleague of mine, and he said I was wrong; he could give me a case in point. The case is so interesting, though it turns out in the end to be understandable, that I thought I'd mention it to show both the dangers and the strengths of the argument of sufficient reason.

Thirty or forty years ago a small industrial firm in Wisconsin—not the great firm in the same business which you all know—made to its profit and surprise a very large alternating-current motor stator and rotor assembly, the biggest thing it had ever made, and worth hundreds of thousands of dollars. It was to be shipped by special flatcar to a Canadian paper mill where it was to be used to run the big pumps or the screens and rolls in the paper mill. The firm surveyed the route, got a special flatcar, and securely placed their big ring in a vertical plane on the flatcar (it weighed about twenty tons), and they sent it off, very happy after months of building it. My informant friend was then a young engineer working on the project. After three days, when the train was first due, they had a telephone call from the paper mill, asking, "Where's our stator; we're anxious to have it?" They said, "We sent it off." "It hasn't got here yet." "Oh, well, these railroads!" After two more days, lots of tracing, no sign of the thing could be found. Finally, the empty flatcar was delivered—with no ring on it. Obviously, the beings from the upper air had stolen the stator! The only evidence was that the lashes were broken, and somehow it had gotten off the car. So my friend was posted to drive along the roads of the states intervening, along the railroad right-of-way, looking for the ring. He did the whole route twice by day, but he never could find the piece at all. There you see the first case known to me of a genuine cargo disappearance by rail. (Theft could be ruled out.)

Twenty years later, as the auto came to replace the railroad, more highways were being built. In the neighborhood of the railroad right-of-way, a considerable swamp was drained to erect highway foundations. There in the mud was the sunken stator. The event was reconstructed: The train had gone around a curve, the lashings had broken, the big ring had rolled off in the middle of the night, down the bank, and sunk into the mud, where nobody saw it again for twenty years. I think this is a partial victory for the faction that says things don't disappear on land. In the sea, on the other hand, they disappear very well. (There's a certain margin to allow for a few wet swampy regions!)

Second, I would like to discuss what I call the "perceptual set," which is a well-known psychological term for the admirable and indispensable art of survival which is built into a human being, in the retina, the visual cortex, the cerebrum, and everywhere else, which tries to spin an orderly set of events out of our very complex percepts. If you try to remember the position of random points, you have a very hard time in doing so; if you let people tell you that these points have a form, you can do it much better. We also have an elimination procedure for making order from the random events of the world. That goes very deep. The trouble is that if you face an unfamiliar set of percepts, you will force them into constructions which will fit, but often at the price of inserting one false link in the chain. That kind of subjective experience is extremely hard to sidestep. I will mention two examples; I submit they will be found very frequently at the bottom of puzzling, earnest, well-informed, and credible eyewitness accounts of strange phenomena.

A state policeman reported a UFO by telephone to some Air Force experts, who knew him and understood his reliability and credibility: "I have myself in sight a flying saucer which has just landed. It is a couple of miles off, blinking on the sun-struck horizon, obviously disk-shaped; a metallic material." After a few hours the experts arrived, and they all gingerly approached the site—to discover an aluminum surplus tank, which was being used to water cattle in the dry range. Everything the policeman had said was right, except for one thing: He had not seen the object land. He was driving on the road for the first time in a couple of weeks, and as he came around a curve he knew very well, he noticed a new strange object in the distance glistening at him. All his other perceptions were correct, but he made the assumption he had seen it land. If any of us examine our own experience, I think we will find, although not perhaps in such an interesting context, such a misapprehension which the "snap of the finger," so to speak, sets right: Of course! That's the effect of perceptual set.

Recently a friend told me of three radio astronomers, himself included, who stood outside Washington, D.C., some years ago watching a large cigar-shaped object in the air, perfectly silent, with visible lighted windows, moving very rapidly past them; independently, they told each other they had each certainly seen the most remarkable kind of unidentified flying object. Suddenly the wind changed, and aircraft engines were heard; the distance adjusted itself, and they recognized they were seeing an ordinary airliner, much nearer than they had thought but not audible because of some peculiar sonic refraction of the wind. A change of perceptual set changed their entire view of the phenomenon.

Third, I'd like to speak about statistical independence. Independent corroborative evidence is necessary, although it is very difficult to obtain in the matter of UFO's because all of our judgments are influenced by the normal spread of the news, as Mr. Sullivan pointed out. This has a direct bearing upon the enormous literature that has been alluded to throughout this symposium. There is in fact a UFO industry; I can call it nothing else. It involves a substantial body of semiprofessional persons, publishers and the like, who derive from this industry an income considerably larger, say, than the whole budget of the AAAS. That's not bad in itself; I have nothing whatever against it. I only say that its presence represents, like the Air Force, on a smaller scale and perhaps even closer to the subject, a vested interest; in the face of which one has to take certain matters into account. Consider George Adamski's notorious UFO book in which, with enormous boldness, he reproduced a Sears Roebuck chicken brooder as his frontispiece. Such a bold photograph is an industrial product—one made with great competence but with no scruples. The book sold more copies than the Condon Report, but it is nevertheless not evidence. Such phenomena—which accompany an industry—make independence hard to pursue and hard to maintain. They therefore reduce the statistical validity of our information, especially when we depend, as we must in matters which involve late-night observations in obscure portions of the world, upon small parties and groups of associates, who must necessarily be regarded as dependent because of the long opportunities they have had to know each others' views and mind.

Fourth, there is the danger of what I call homogenization: that is, the placing into one category of disparate objects, where the categorical resemblance is itself an assumption. Disk-shaped, cigar-shaped: there is no question that an effort to characterize the whole complex field of vision by a few metaphors of shape represents one difficulty.

Fifth, of course there are real phenomena we do not understand. I venture to say that there will never be a time when an AAAS meeting can be held where all phenomena will be understood in any topic under discussion. At least, I hope that time will not come! There are many phenomena still to understand.

One example, which I think is a spectacular one, is revealed in a photograph published by M.G.J. Minnaert, an expert on atmospheric and meteorological optics.[2] It was taken by a woman-passenger on a liner in the Indian Ocean, at the request of the twenty or thirty other passengers in the first-class salon, who all testified that they had seen this very same phenomenon. They got the woman, who was the best photographer among them, to walk out to the rail of the ship in clear air to take the photograph. It shows the sun at low altitude, doubled in a very curious way. In the picture are two suns, each precisely at the same altitude over the horizon, with no differences in color, shape, or size. As far as I know, no one has been able to explain this phenomenon. You could simulate the effect if you were to support, a thousand yards from the ship, a square of flat plastic, say, some tens of yards on an edge, perfectly vertical within half a degree or so, tilted within a certain broad range from the ship's course: this would yield a proper bright reflected image. It's very unlikely that any looming or atmospheric mirage phenomenon could produce this kind of effect, so precisely on the sun's level, so precisely the sun's size. And the photograph is hard evidence. No one suggests that this is a brand new phenomenon. It is simply something we don't understand about the propagation of light in the atmosphere that hour. A small modification in propagation, but of the right sort, would do, for the sun is bright enough to yield a good image even if most light is wasted.

This is quite a striking case. It is relevant to what Professor McDonald has to say (Chapter 5), because of the objectivity of his most interesting cases with radar aircraft (more interesting, I would say, than many previous accounts I have heard). Perhaps the time has not yet come when we can do this, but we need eventually an accounting from high in the military hierarchy, concerning more of the farflung systems that organization operates. We need to know more about unusual drones and deliberate "spoofing" of all kinds. We need theoretical studies, and these will be difficult, on double radar returns, and other phenomena. I find the events puzzling, but I don't know enough about the complicated systems to draw a link-by-link inference.

Sixth: It is plain to me (this also was alluded to by Carl Sagan) that there is a whiff of paranoia which is perhaps characteristic of our terrible times, not by any means only in the literature that supports the extraterrestrial hypothesis—the UFO literature—but also among the Air Force personnel and the other people that seek to refute UFO evidence. My best case for this paranoia is Appendix U to the Condon Report, which is a report of the Robertson Panel, which had met sixteen years earlier. The only thing that has been excised from the original report, as far as I can tell, is the phrase "the Director"—not only on one occasion, but on several occasions. There must be in the Central Intelligence Agency a specific shibboleth that says we never refer to "the Director" (or maybe "the Vice-Director") as present at meetings, or in published literature. (I don't know the regulation; I just reconstruct it from the evidence.) This might be a wrong inference; if so, I shall stand corrected. But there can be little doubt that the mere presence of the director of the Central Intelligence Agency, or his vice-director, at a meeting sixteen years old was kept out of the record. The Condon Report had to accept these deletions because they couldn't get them easily reinstated via the bureaucratic chain. It's a small demonstration of a general institutional paranoia.

The whole emphasis on secrecy, on the risk of overloading the report systems, on fear, which was in the 1950's characteristic of the entire Air Force organization—especially of the Strategic Air Command—has had a lot to do with the persistence of American support for UFO's. If you ask, "Should we be suspicious that the Air Force may have treated these data badly, in view of their general record of what they did with other things?" I say, yes, we should be suspicious. If the Air Force presented data in this way in a court of law, sharp attorneys would make monkeys of them. I don't believe that there is the slightest evidence of Air Force suppression of data in this whole affair, but I can well understand persons who believe there is. The rational case for the existence of supression here is stronger, probably, than my firm belief to the contrary. So there is around this whole story an unhappy atmosphere which we must take into account. That's a special case of a more general point made by Professor Sagan, who began most admirably by saying that in dealing with this question we must be especially careful to guard against errors in both directions.

I have now, after a couple of years of fairly systematic listening and reading, no sympathy left for the extraterrestrial hypothesis. That does not mean that I know what is going on; I don't, at least not in every area. But the apparent strength of the evidence is not such as to make me regard the UFO phenomena a matter of very high priority for myself. I have no objection to what others do. I would not, for example, support any substantial federal investigation of these matters, but I would always be sympathetic to any positive effort to follow a link-by-link evidential chain to produce any kind of information, whether from the domain of science, or the domain of hearsay, or from any source. Any person who tries to make order out of his perceptions will always, I hope, have my ear and my support, and he should have at his disposal the investigative facilities which could reasonably be granted him. I think that it is demonstrably wrong to take the attitude that many take in saying that science has nothing to do with it; science is cognizant of the entire world.

There passed over my desk a year ago a reprint from Edwardian England, a book of photographs, showing rather elegant scenes—drawing rooms, gardens, and so on. In every one of these little scenes there was, somewhere in the photograph, a six-inch-high young woman with a simple veiling over her nude form, and with two (or sometimes four) small dragonfly-like wings attached to her. Photographs: excellent hard data! Unfortunately my skeptical training is so strong that, in spite of having seen these fifty good photographs, I am unconvinced that there were real ladies like that—very like the lady on the White Rock bottle, but only six inches high, sitting around in Edwardian drawing rooms with London gentlemen—I will remain unconvinced, until a link-by-link study of some evidential case has convinced me that there is no link missing, and that the chain of evidence is very strong. For I have another hypothesis!

On the other hand, I'm most concerned that the people who pursue these items, who see strange things in the sky in our present context, will not conceal it, will not take the view that they will be ridiculed if they announce their experience. On the contrary, they should try to come to grips with experience; they should ask themselves: How do I know how big something is when I see it only as a mark in the sky? When a child reaches for the moon, that is a perfectly reasonable interpretation of a half-degree object. You can almost reach a quarter—you can reach a dime, anyway—if it looks the size of the moon. The educational side of this appeal is to try to get people to understand that we humans do not immediately perceive the world as it is; rather, we are elaborate computers with an enormous preset routine and much programing, both genetic and cultural; and we have to interpret all the data we get. That interpretation, *whatever it is,* is subject to error. I think this public appeal is an important part of the task of science; I hope it will continue to be so.

When at last I shall see a case with the necessary properties—multiple, independent, link-by-link, verified chains of evidence—I will apply to the reporter of that evidence the judgment with which Biot ends his narrative: "I count myself happy if it turns out that it was I who put beyond doubt one of the most astonishing phenomena which men have ever observed."

## NOTES

1. Jean-Baptiste Biot, *Mémoires de l'Institut de France* 7 (1806): 224.
2. M. G. J. Minnaert, *Journal of the Optical Society of America* 58 (1968): 297.

# Addendum: Discussion

REPORTED BY THE EDITORS

[A discussion followed the presentation of the various papers at the symposium. It included questions and comments by those who presented the papers and by others in attendance. This summary was prepared by the Editors.]

Dr. Sagan expressed some doubts about Dr. Roach's "Lilliput" hypothesis (Chapter 3). He stressed that any ecosystem on a Lilliputian star is ultimately tied to a heat engine, such as that provided by green-plant photosynthesis on the earth. The optimal efficiency of such an engine is $1-(T_2/T_1)$, where $T_2$ is the temperature of the sink in the heat engine and $T_1$ is the temperature of the source. In the case of the earth, the source is the solar photosphere, which is very much hotter than the earth's surface, which is the sink, and the efficiency can be quite high. In the case of a Lilliputian star there is no sun in the sky and the efficiency will be exceedingly low. Dr. Sagan doubts whether any form of life, much less intelligent life, could exist on a Lilliputian star.

Dr. Page asked Dr. Hall whether the range and direction of hysterical contagion (Chapter 8) might depend on the subject of the hysteria; e.g., flying saucers are seen to have greater range than June bugs and might be expected to endure longer than a swarm of June bugs. Dr. Hall replied that self-corrective processes generally account for the limits on hysterical contagion, but that all efforts to introduce self-correction on UFO's (study groups, symposia, and so forth) have failed to limit the waves of

UFO sightings, which spread all over the world, and have not halted them in twenty-two years. Dr. Price-Williams pointed out that the term "UFO" includes such a variety of reported sightings and events that it is almost impossible to "debunk" all of them.

Drs. Sagan and Baker mentioned that the space-radar surveillance systems of the Air Force and the Navy would be ideal tools for detecting and examining high-altitude UFO's. But apparently all unknowns not on "interesting" trajectories (such as ballistic intercontinental trajectories or orbital ellipses) are not noted or stored by the computer system. The rejects of these radar systems, which must be extraordinarily sensitive, are of major interest for the study of UFO's. The prospect of utilizing such an operational system for the investigations of UFO's, however, seems small. Dr. Morrison raised the possibility that some of the enigmatic cases cited by Dr. McDonald (Chapter 5)—especially combined radar and visual sightings—may be the result of "spoofing"; that is, the deliberate penetration of U.S. airspace, as a means of testing defense readiness, either by an arm of the U.S. Air Force itself or by foreign powers. Dr. Sagan pointed out that both of these areas, in addition to others reported by the Robertson Panel (1953), indicated a possible conflict between solution of UFO problems and Department of Defense interests, and may keep details of some interesting cases from being fully revealed.

It was asked whether the AAAS should take some action if further evidence demonstrates that some of the UFO's are extraterrestrial. There were no AAAS officials present, but several of the invited speakers suggested that the AAAS should, under these circumstances, organize further study of the data.

Another question was whether the invited speakers were thoroughly familiar with data on UFO's. Dr. Sagan answered that some had more familiarity than others, and that this was clear from their remarks. Obviously, the radar experts had less contact with visual sightings than Dr. Hynek, who has spent much of his time in the last twenty years reviewing UFO sightings. Dr. Page remarked that *no* one man has read *all* the UFO literature, since even the Catoe bibliography (Chapter 1, note 1) is not complete.

A further question noted Dr. Menzel's reference to N rays (see p. 136) and pointed out that some scientists had been skeptical of X rays when they were first discovered. Dr. Page recognized this as a basic issue mentioned by several of the participants. Like many philosophers, he feels that science benefits from both stability and change. When new observations are made, scientists first try to "explain" them in terms of the current theoretical structure—and usually they succeed. But every now and then such an explanation seems impossible, and someone proposes a new theory, a classic case being Einstein's Special Relativity. For a while there is proper skepticism of such a theory and proper efforts to prove it wrong. A good scientist, as Walter Sullivan says (Chapter 13), should exploit the power of organized knowledge in current theory, but should keep his mind open to possible changes.

Another basic question was how one can tell whether distant objects seen, or photographed, or detected by radar are "real solid objects," or transients, or illusions. Dr. Page admitted that this decision is difficult; we usually accept explanations that have high probability of matching all the observations but we can never be sure that some other, uninvented explanation or theory will not fit the data as well or better. Thus a moving "object" may simply be a searchlight beam shining on a cloud layer, but it might also be a self-luminous aircraft. As Dr. Hardy showed (Chapter 7) radar returns may come from clear-air turbulence, or from clouds of air with higher temperature, higher humidity, or higher ionization, than the surrounding atmosphere. Hence a radar echo is not evidence of a "solid object." The combination of visual and radar detections in the same direction is very likely to indicate a solid body, but it could be an auroral display. There is no one certain criterion for "real solid objects" at a distance (except collision with a known object such as an airplane or bullet), but the combination of several simultaneous observations make such identifications very probable.

A questioner showed two slides, one a photograph, the other the witnesses' drawing of the same UFO, and asked whether the "fuzziness" of the photo (in marked contrast to the sharp image described by the witnesses) could be due to the different colors of light detected by the eye and the camera. Dr. Page answered that this is unlikely, since most camera lenses are designed to focus all colors of visual light sharply. He referred to analyses of photographs by NICAP and the Colorado Project, which made full use of known lens and photographic defects such as internal reflections and the "spreading" of overexposed images. The questioner expressed his belief that some UFO's come from "another dimension."

Another questioner referred to recent physiological studies of hallucinations produced by strong magnetic fields across the subject's brain. The question was: "Can ball lightning and other plasma phenomena produce hallucinations in witnesses and thus account for some unexplained reports?" Walter Sullivan replied that, although he is no expert in the physiology of magnetohallucination, the magnetic fields of ball lightning and other electrical discharges at distances of 100 feet or more would be very much weaker than those used in laboratory experiments. Dr. Sagan added that, even if these weak magnetic fields could produce hallucinations, they would not account for several witnesses sharing the same hallucination.

Another questioner asked whether experienced radar operators could distinguish between real targets and "bogies," and Dr. McDonald answered that under most circumstances they can. The Chairman asked Dr. McDonald whether he would care to reply to Dr. Menzel's criticism (see p. 130) and he said that a full reply would take more than the time available. Individual cases require a great deal more detailed discussion. For instance, the Salt Lake City case in 1961, which gets a bare fourteen lines as a Selected UFO Case on p. xxx, was identified by Dr. Menzel as a sundog (see p. 134) without, in Dr. McDonald's view, regard for well-established facts reported by six witnesses who used binoculars from the ground, as well as the aircraft-pilot witness. All witnesses agree that the object was at first low on the horizon, seen in front of a distant mountain, and that it suddenly "took off" at high speed in a "steep climb." This scarcely fits the USAF identification with the planet Venus, known to be in the southeastern sky. Since the sun was high on the meridian (due south), Menzel's "sundog" explanation cannot fit the reported direction (azimuth 170° from the Utah airport) or the rapid motion. Dr. McDonald felt that one case like this can prove little by itself, but he stressed the fact that there are many more where Dr. Menzel and USAF Project Bluebook have accepted explanations that simply do not account for the significant and well-established aspects of witnesses' reports. He thus concludes that atmospheric phenomena cannot account for these cases; they are "true" UFO's.

The Chairman invited comments by Dr. George Kocher, astronomer at the University of Southern California, who provided the summary of the Newton, Illinois, 1966 case (see Selected UFO Cases, p. xxix). Dr. Kocher said that he became interested in the extraterrestrial hypothesis several years ago, and he interviewed the Newton witnesses to satisfy his curiosity about the likelihood of sighting extraterrestrial spacecraft. He feels that this possibility has major implications in several scientific fields, and noted that most of the scientists who have spent the necessary time to interview witnesses and analyze a case thoroughly conclude (as he does) that the possibility of extraterrestrial visits is worthy of further investigation.

Dr. Kocher went on to say that those scientists interested in further investigation would all like to use the data collected by the USAF Project Bluebook over the past twenty-two years, but that this is difficult because the USAF has classified, as "confidential" or higher, most of the interesting cases. Now that the project is discontinued, Dr. Kocher hopes that the Air Force will declassify the files and make them available for scientific study. Dr. Page replied that several of the invited speakers had already discussed action to insure preservation of the Bluebook files. He had phoned Dr. Marr at the USAF Archives at Maxwell AFB, Montgomery, Alabama, and was told that "no classified papers on UFO's will be stored in the Archives." This is considered to leave open the possibility that the USAF will destroy the classified UFO reports. Most of the speakers agreed to write a joint letter to the Secretary of the Air Force requesting his cooperation in preserving the UFO data, and the following letter was mailed to Mr. Robert Seamans, Jr., on December 30, 1969:

> The scientists listed below, convened at a General Symposium during the Annual Meeting of the Association, understand that USAF Project Bluebook has been discontinued in accordance with Dr. E. U. Condon's recommendation in the Colorado Study of Unidentified Flying Objects. We know that Project Bluebook accumulated, over the past two decades, irreplaceable data of great historical interest and potential value to physical and (particularly) behavioral scientists.

After two days' discussion of the data involved, the Colorado Study, and several proposed studies by sociologists and psychologists, we formally request that you, Mr. Secretary

(1) insure that *all* of the material, both classified and unclassified, be preserved without alteration or loss,
(2) declassify promptly all documents filed by the Aerial Phenomena Section of the Wright-Patterson Air Force Base which are classified by virtue of AFR 200-17 and AFR 80-17,
(3) make all the unclassified documents available to qualified scientific investigators at a more suitable location than the USAF Archives (we recommend a major university in the Midwest), and
(4) order an annual review of the remaining classified documents in the present file to determine when they can be declassified without alteration in accordance with current USAF security procedure.

My twelve colleagues, who receive copies of this letter, would appreciate your favoring us with a reply. I can distribute it to the others if you address it to Dr. Page, 18639 Point Lookout Drive, Houston, Texas 77058.

Sincerely,

Thornton Page (Wesleyan University)
Chairman, AAAS Special Committee, for

Walter Orr Roberts, Retiring President, AAAS
Franklin E. Roach, University of Hawaii
William Hartmann, University of Arizona
Lester Grinspoon, Harvard University
Robert Hall, University of Illinois
Philip Morrison, Massachusetts Institute of Technology
Douglass Price-Williams, Rice University
J. Allen Hynek, Northwestern University
James McDonald, University of Arizona
Carl Sagan, Cornell University
Walter Sullivan, *The New York Times*
George Kocher, University of Southern California

## Afterword

The following unsigned, undated form letter was sent by the Secretary's office to Chairman Page:

DEPARTMENT OF THE AIR FORCE
Washington 20330

OFFICE OF THE SECRETARY

We wish to acknowledge receipt of your recent inquiry. Please accept this form of response so that we may give you a reply without undue delay.

On December 17, 1969 the Secretary of the Air Force announced the termination of Project Blue Book, the Air Force Program for the investigation of unidentified flying objects (UFOs).

The decision to discontinue UFO investigations was based on an evaluation of a report prepared by the University of Colorado entitled, "Scientific Study of Unidentified Flying Objects;" a review of the University of Colorado's report by the National Academy of Sciences; past UFO studies; and Air Force experience investigating UFO reports during the past two decades.

As a result of these investigations and studies, and experience gained from investigating UFO reports since 1948, the conclusions of Project Blue Book are: (1) no UFO reported, investigated, and evaluated by the Air Force has ever given any indication of threat to our national security; (2) there has been no evidence submitted to or discovered by the Air Force that sightings categorized as "unidentified" represent technological developments or principles beyond the range of presentday scientific knowledge; and (3) there has been no evidence indicating that sightings categorized as "unidentified" are extraterrestrial vehicles.

With the termination of Project Blue Book, the Air Force regulation establishing and controlling the program for investigating and analyzing UFOs has been rescinded, and Project Blue Book records have been transferred to the Air Force Archives.

Attached for your information is the Project Blue Book sighting summary for the period 1947–1969. Also included is a listing of UFO-related materials currently available through publication outlets of the federal government.

Your interest in the United States Air Force is appreciated.

Sincerely,

JAMES H. AIKMAN, Lt Colonel, USAF
Chief, Civil Branch
Community Relations Division
Office of Information

Total UFO Sightings, 1947–1969

| Year | Total sightings | Unidentified |
|---|---|---|
| 1947 | 122 | 12 |
| 1948 | 156 | 7 |
| 1949 | 186 | 22 |
| 1950 | 210 | 27 |
| 1951 | 169 | 22 |
| 1952 | 1,501 | 303 |
| 1953 | 509 | 42 |
| 1954 | 487 | 46 |
| 1955 | 545 | 24 |
| 1956 | 670 | 14 |
| 1957 | 1,006 | 14 |
| 1958 | 627 | 10 |
| 1959 | 390 | 12 |
| 1960 | 557 | 14 |
| 1961 | 591 | 13 |
| 1962 | 474 | 15 |
| 1963 | 399 | 14 |
| 1964 | 562 | 19 |
| 1965 | 887 | 16 |
| 1966 | 1,112 | 32 |
| 1967 | 937 | 19 |
| 1968 | 375 | 3 |
| 1969 | 146 | 1 |
| Total | 12,618 | 701 |

## UFO Materials

*UFOs and Related Subjects: An Annotated Bibliography*. Lynn E. Catoe. Prepared by the Library of Congress Science and Technology Division. Library of Congress Card Catalog No. 68-62196. For sale by the Superintendent of Documents, U.S. Government Printing Office, Washington, D.C. 20402, $3.50. GPO #D301.45-19-2:68-1656.

*Aids to Identification of Flying Objects*. For sale by the Superintendent of Documents, U.S. Government Printing Office, Washington, D.C. 20402, 36 p., 20¢ per pamphlet. GPO #D301.2:F67.

*Scientific Study of Unidentified Flying Objects*. Study conducted by the University of Colorado under contract F44620-67-C-0035. Three volumes, 1,465 p. 68 plates. Photoduplicated hard copies of the official report may be ordered for $3.00 per volume, $9.00 the set of three, as AD 680 975, AD 680 976, and AD 680 977, from the Clearinghouse for Federal Scientific and Technical Information, U.S. Department of Commerce, Springfield, VA 22151.

*Review of University of Colorado Report on Unidentified Flying Objects.* Review of report by a panel of the National Academy of Sciences. National Academy of Sciences, 1969. 6 p. Photoduplicated hard copies may be ordered for $3.00 as AD 688 541 from the Clearinghouse of Federal Scientific and Technical Information, U.S. Department of Commerce, Springfield, VA 22151.

The last-named review by the National Academy Panel was described in the Academy's news report for February 1969 in which the panel referred to papers by James E. McDonald and Donald H. Menzel used in its review. The panel, chaired by G. M. Clemence of Yale, concluded that the scope of the Condon Report was adequate, the methodology well chosen, and the conclusions justified. It concurred with E. U. Condon's recommendation that Project Blue Book be terminated, although there are important areas of atmospheric research that should be continued both by individual scientists and such government agencies as the Environmental Science Services Administration, the National Aeronautic and Space Administration, and the National Center for Atmospheric Research. However, "the study of UFO's in general is not a promising way to expand scientific understanding of the phenomena." The panel report has been reprinted, e.g., in *Icarus 11:* 440–443. There was unanimous concurrence of the eleven panel members.

After a phone call from Page, the Community Relations Information Officer (SAFOI) in the Office of the Secretary of the Air Force followed up the form-letter response with the following note:

DEPARTMENT OF THE AIR FORCE
Washington 20330

OFFICE OF THE SECRETARY

13 Jan 1970

Dear Dr. Page:

This is in reply to your recent letter and our telephone conversation this date.

Please be advised that no Project Blue Book records will be destroyed. There are no longer any classified UFO records. All have been declassified in accordance with Air Force directives.

Additionally bona fide researchers and news media representatives will be granted access to the records upon application to HQ USAF (SAFOI), The Pentagon, Washington, D.C. 20330.

I trust this information will be helpful and if you have any further questions regarding this matter, please let me know.

Sincerely,

S/James H. Aikman

Lt Colonel, USAF
Chief, Civil Branch
Community Relations Division
Office of Information

In the meantime, Donald Menzel had written to the Secretary and received the following personal reply:

DEPARTMENT OF THE AIR FORCE
Washington 20330

OFFICE OF THE SECRETARY

Jan 26, 1970

Dear Donald:

I appreciate your letters, which were awaiting me upon my return from Southeast Asia, and am glad to know that I have your support in my decision to discontinue UFO investigations.

Although our news release contained no directives to Air Force bases concerning the cessation of local UFO investigations and providing information to private UFO groups, we did provide our bases with guidance in these areas. Specifically, should an individual report a UFO sighting to an Air Force base, the base information officer will suggest that the local police department be consulted. If the individual feels that something of scientific importance exists, the information officer will suggest that a responsible member of the scientific community be contacted. However, the local air base will check to determine whether the base has seen and identified the source, but will not conduct a formal investigation.

We have no intention of providing private UFO investigative organizations with special information on UFO sightings reported to our air bases.

For your information, we received a formal proposal from one such organization requesting that we refer sighting reports to several established private UFO groups. We declined the proposal.

After careful consideration we decided to maintain Project Blue Book records intact in the Air Force Archives at Maxwell Air Force Base, Alabama. Access to these records, which are all unclassified, will be granted to bona fide news media representatives and researchers upon approval of the Secretary of the Air Force, Office of Information (SAFOI).

Your interest and views in this matter are greatly appreciated and I trust that the above clarifies the Air Force position on this subject.

Sincerely,

S/Robert C. Seamans, Jr.

Somewhat later, in 1970, James McDonald was able to see, after some difficulty, specified reports in the Project Blue Book files at Maxwell Air Force Base in Alabama. He found that he could see only items that he named in advance, and waited for copies to be made in which all individuals' names were deleted (scarcely convenient for any serious study). J. Allen Hynek has suggested that a full set of copies with all names deleted should be made and housed in a library at some more convenient location, where sociologists and atmospheric scientists could have easy access. This might be an expensive operation, and has not yet been undertaken (September 1972).

The reluctance of the U.S. Air Force to continue collection of UFO reports can be understood (since it is generally agreed that UFO's are not a threat to national security), but the reluctance to allow research of old records has been challenged by many of the younger scientists, particularly in the fields of sociology and psychology.

One of the interesting psychological phenomena uncovered in the course of arranging the AAAS symposium on UFO's in December 1969 was the strong opposition of several older physical scientists. These men were convinced that AAAS sponsorship would somehow lend credence to "unscientific" ideas, despite the plans for discussion of all sides of the question, as described in the Introduction, in the traditional manner of science. Letters demanding the cancellation of the symposium were written to the AAAS Board where Walter Orr Roberts argued the Association's duty to promote scientific discussion of any topic considered significant by so many of its members. At one point the older scientists' letters swung a vote of the AAAS Section D (Astronomy) to refuse its sponsorship of the UFO symposium; later letters were sent to congressmen and even to the Vice President of the United States urging their intervention to secure cancellation of the symposium.

The actual outcome of the symposium—the papers published in this volume, and the discussion—has served, we hope, to improve the understanding of the UFO phenomena, by scientists and laymen alike.

Carl Sagan
Thornton Page

# Index